# Minitab Manual

## for

## Moore and McCabe's

## Introduction to the Practice of Statistics
*Third Edition*

**Michael Evans**
*University of Toronto*

W. H. Freeman and Company
New York

Printed in the United States of America


ISBN: 0-7167-2785-4


First printing 1999, Hamilton

# Contents

# Preface

This Minitab manual is to be used as an accompaniment to *Introduction to the Practice of Statistics*, Third Edition, by David S. Moore and George P. McCabe, and to the CD-ROM that accompanies this text. We abbreviate the textbook title as IPS.

Minitab is a statistical software package that was designed especially for the teaching of introductory statistics courses. It is our view that an easy-to-use statistical software package is a vital and significant component of such a course. This permits the student to focus on statistical concepts and thinking rather than computations or the learning of a statistical package. The main aim of any introductory statistics course should always be the "why" of statistics rather than technical details that do little to stimulate the majority of students or, in our opinion, do little to reinforce the key concepts. IPS succeeds admirably in communicating the important basic foundations of statistical thinking, and it is hoped that this manual serves as a useful adjunct to the text.

It is natural to ask why Minitab is advocated for the course. In the author's experience, ease of learning and use are the salient features of the package, with obvious benefits to the student and to the instructor, who can relegate many details to the software. While more sophisticated packages are necessary for higher level professional work, it is our experience that attempting to teach them in a course forces too much attention on technical aspects. The time students need to spend to learn Minitab is relatively small and that it is a great virtue. Further Minitab will serve as a perfectly adequate tool for many of the statistical problems students will encounter in their undergraduate education.

The manual is divided into two parts. Part I is an introduction that provides the necessary details to start using Minitab and in particular how to use worksheets. Not all the material in Part I needs to be absorbed on first reading. We recommend reading I.1-I.10 before starting to use Minitab. The material in I.11 is more for reference and for later reading. References are made to these sections later in the manual and can provide the stimulus to read them. Overall the introductory Part I also serves as a reference for most of the nonstatistical commands in Minitab.

Part II follows the structure of the textbook. Each chapter is titled and numbered as in IPS. The last two chapters are not in IPS but correspond to optional material included on the CD-ROM. The Minitab commands relevant to doing the problems in each IPS chapter are introduced and their use illustrated. Each chapter concludes with a set of exercises, some of which are modifications of or related to problems in IPS and many of which are new and specifically designed to ensure that the relevant Minitab material has been understood. There are also appendices dealing with some more advanced features of Minitab, such as programming in Minitab and matrix algebra.

Minitab is available in a variety of versions and for different types of computing systems. In writing the manual we have used Version 12 for Windows, as discussed in the references in Appendix F, but have tried to make the contents of the manual compatible with earlier versions and for versions running under other operating systems. We have tried to point out where differences exist in earlier versions. The core of the manual is a discussion of the session commands while not neglecting to refer to the menu commands and the superior plotting features available with more recent versions of Minitab. Overall we feel that the manual can be successfully used with most versions of Minitab.

This manual does not attempt a complete coverage of Minitab. Rather we introduce and discuss those concepts in Minitab that we feel are most relevant for a student studying introductory statistics with IPS. We do introduce some concepts that are strictly speaking not necessary for solving the problems in IPS where we feel that they were likely to prove useful in a large number of data analysis problems encountered outside the classroom. While the manual's primary goal is to teach Minitab, more generally we want to help develop strong data analytic skills in conjunction with the text and the CD-ROM.

Thanks to Patrick Farace and Chris Spavens of W.H. Freeman and Com-

pany for their help and consideration. Also thanks to Rosemary and Heather.
For further information on Minitab software contact:

Minitab Inc.
3081 Enterprise Drive
State College, PA 16801 USA
ph: 814.328.3280
fax: 814.238.4383
email: Info@minitab.com
URL: http://www.minitab.com

# Part I

# Minitab for Data Management

**New Minitab commands discussed in this part**

| | | | | | |
|---|---|---|---|---|---|
| code | concatenate | convert | copy | delete | erase |
| exit | help | info | insert | journal | let |
| name | outfile | print | rank | read | restart |
| retrieve | save | set | sort | stack | unstack |
| write | | | | | |

# 1 Manual Overview and Conventions

The manual is divided into two parts. Part I is concerned with getting data into and out of Minitab and giving you the tools necessary to perform various elementary operations on the data so that it is in a form in which you can carry out a statistical analysis. You don't need to understand everything in Part I to begin doing the problems in your course. Part II is concerned with the statistical analysis of the data set and the Minitab commands to do this. The chapters in Part II follow the chapters in *Introduction to the Practice of Statistics* (IPS hereafter) and are numbered accordingly. Before you start on Chapter II.1, however, you should read I.1-I.10 and leave I.11 for later reading.

Minitab is a software package that runs on a variety of different types of computers and comes in a number of versions. This manual does not try to describe all the possible implementations or the full extent of the package. We limit our discussion to those features common to most versions of Minitab. Also we present only those aspects of Minitab relevant to carrying out the statistical analyses discussed in IPS. Of course this is a fairly wide range of analyses but the full power of Minitab is not necessary. Depending on the version of Minitab you are using there may be many more useful features and we encourage you to learn and use them. Throughout the manual we point out what some of the additional useful features of Minitab are and how you can go about learning how to use them. Version 12 refers to the most current version of Minitab at the time of writing this manual. The material in this manual should be enough, however, to successfully carry out in any version of Minitab the analyses required in your course.

In this manual special statistical or Minitab concepts will be highlighted in *italic* font. You should be sure that you understand these concepts. We will provide a brief explanation for any terms not defined in IPS. When a reference is made to a Minitab *session command* or *subcommand* its name will be in **bold** font. At times we will also refer to *menu commands* that are available in some versions of Minitab. Menu commands are accessed by clicking the left button of the mouse on items in lists. We use a special notation for menu commands. For example,

A ▶ B ▶ C

is to be interpreted as left click the command A on the menu bar, then in the list which drops down, left click the command B and finally left click C. The menu commands will be denoted in ordinary font exactly as they appear. The record of a Minitab session – the commands we type and the output obtained – will be denoted in `typewriter` font as will the names of any files used by Minitab, variables, constants and worksheets.

At the end of each chapter we provide a few exercises that can be used to make sure you have understood the material. We recommend, however, that whenever possible you use Minitab to do the problems in IPS. While many problems can be done by hand you will save a considerable amount of time and avoid errors by learning to use Minitab effectively. We also recommend that you try out the Minitab commands as you read about them as this will ensure full understanding.

# 2 Accessing and Exiting Minitab

The first thing you should do is find out how to access the Minitab package for your course. This information will come from your instructor, system personnel or from your software documentation if you have purchased Minitab to run on your own computer.

In some cases this may mean you type a command such as **minitab** at a computer system prompt, and hit the Enter or Return key on the keyboard, after you have logged on, i.e. provided a login name and password to the computer system being used in your course. Typically you will see the prompt

    MTB >

on your screen and this indicates that you have started a Minitab *session*. Hereafter we refer to this as the *session environment*.

In other cases – e.g. Minitab for Windows has been installed on your own computer – you may double click an icon that corresponds to the Minitab program or in Windows 95 use the Start button and click on Minitab in the Programs list. In this case the program opens with a number of overlaid windows all lying within a *Minitab window*. Left clicking the mouse anywhere on a particular window brings that window to the foreground, i.e. makes it the *active* window. For example, clicking in the *Session window* will bring up a window containing the MTB > prompt. Alternatively you can use the command Window ▶ Session in the *menu bar* at the top of the Minitab window to bring this window to the foreground. We refer to this implementation of Minitab hereafter as the *window environment*. In fact, for this manual, the window environment is exactly the same as the session environment provided you stay within the Session window. If you are using Version 12 of Minitab you may not see the MTB > prompt in your Session window and for this manual it is important that you do so. In this case the menu command Editor ▶ Enable Command Language in the menu bar at the top of the Minitab window makes this prompt appear in the Session window. You can ensure that this prompt always appears in your Session window for this version of Minitab using Edit ▶ Preferences ▶ Session Window ▶ Select ▶ Command Language Enable ▶ OK.

While most of our discussion is focused on the session environment there are some conveniences to the window environment that you will want to take advantage of, if it is available to you, and we will point these out throughout the manual. If you do not have access to the window environment then you can just ignore any subsequent references we make to this. Also even if you have the window environment you can ignore any such references and do all your work in the Session window.

In the session environment Minitab *commands* are typed after the MTB > prompt and executed when you hit the Enter or Return key. For example, the first command you should learn is **exit**, as this takes you out of your Minitab session and returns you to the system prompt or operating system. In the window environment you can also access the **exit** command using File ▶ E̲xit. In some implementations this may instead be the command **stop**.

Minitab is an interactive program. By this we mean you supply Minitab with input data, or tell Minitab where your input data is, and when you give a command to Minitab telling it to do something to your data, Minitab responds instantaneously to your command. You are then ready to give another command. It is also possible to run a collection of Minitab commands in a batch program; i.e. several Minitab commands are executed sequentially before the output is returned to the user. The batch version is useful when there is an extensive number of computations to be carried out. This is discussed in Appendix C and can be read after you are more familiar with Minitab.

# 3 Files Used by Minitab

Minitab can accept input from a variety of files and write output to a variety of files. These files are distinguished by a *file name* and an *extension* that indicates the type of file it is. For example, `marks.mtw` is the name of a file that would be referred to as 'marks' (note the single quotes around the file name) within Minitab. The extension `.mtw` indicates that this is a Minitab worksheet. We describe what a worksheet is in I.5. This file is stored somewhere on the hard drive of a computer as a file called `marks.mtw`.

There are other files that you will want to access from outside Minitab, perhaps to print them out on a printer. Depending on the version of Minitab you are using, to do this you may have to exit Minitab and give the relevant system print command together with the full path name of the file you wish to print. As various implementations of Minitab differ as to where these files are stored on the hard drive you will have to determine this information from your instructor or documentation or systems person. For example, in the windows environment the full path name of the file could be

`c:/program files/mtbwin/marks.mtw`

or something similar. This path name indicates that the file `marks.mtw` is stored on the C hard drive in the directory called `program files/mtbwin`. We will discuss several different types of files in this chapter.

In many versions of Minitab there are restrictions on file names. A file name can be at most 8 characters in length using any symbols except # and ' and the first character cannot be a blank. The length restriction on file names is removed in Version 12. It is generally best to name your files so that the file name reflects its contents. For example, the file name `marks` may refer to a data set composed of student marks in a number of courses.

# 4 Getting Help

At times you may want more information about a command or some other aspect of Minitab than this manual provides or you may wish to remind yourself of some detail that you have partially forgotten. Minitab contains an online manual which is very convenient. From the `MTB >` prompt you can use the **help** command for this purpose. Typing **help** followed by the name of the command of interest, and hitting Enter, will cause Minitab to produce relevant output. The output will vary depending on the version of Minitab you are using. In some versions a separate window will pop up containing the information while in others this will be printed directly onto your screen. For example, asking for help on the command **help** itself via the command

```
MTB >help help
```

will give you an overview of what help information can be accessed on your system.

In the window environment you can access this information directly via clicking on Help in the Menu bar and then using the table of Contents or using a Search.

The depth and extent of the information obtained in this way varies considerably by implementation of Minitab. In the window environment this manual is very complete and we will often refer you there for more details concerning some topic. For the session environment, however, we try to give a fairly complete discussion of all the relevant concepts.

# 5 The Worksheet

The basic structural component of Minitab is the *worksheet*. Basically the worksheet can be thought of as a big rectangular array, or matrix, of *cells* organized into rows and columns. Each cell holds one piece of data. This piece of data could be a number, i.e. *numeric data*, or it could be a sequence of characters, such as a word or an arbitrary sequence of letters and numbers, i.e. *text data* (also referred to as *alpha data* in some versions of Minitab). Data often comes as numbers such as 1.7, 2.3, ... but sometimes it comes in the form of a sequence of characters such as black, brown, red, etc. Typically sequences of characters are used as identifiers in classifications for some variable of interest, e.g. color, sex. A piece of text data can be up to 80 characters in length in Minitab. Version 12 also allows for *date data* which is data especially formatted to indicated a date as in 3/4/97. We will not discuss date data.

If possible try to avoid using text data with Minitab, i.e. make sure all the values of a variable are numbers, as dealing with text data in Minitab is more difficult. For example, denote colors by numbers rather than by names. Still there will be applications where data comes to you as text data, e.g. in a computer file, and it is too extensive to convert to numeric data. So we will discuss how to input text data into a Minitab worksheet but we recommend that in such cases you use the **convert** command, discussed in I.11.5, to convert this to numeric data. In Version 12 of Minitab it is somewhat easier to deal with text data than earlier versions and this proviso is not as necessary.

Table 1 is a good way to visualize a worksheet. In fact if you are using Minitab in the window environment there is a *Data window* which looks very similar to this. Notice that the columns are labeled C1, C2, etc. and the rows are labeled 1, 2, 3, etc. We will refer to the worksheet depicted in Table 1 as the `marks` worksheet hereafter and will use it throughout Part I to illustrate various Minitab commands.

Data arises from the process of taking measurements of variables in some real-world context. For example, in a population of students suppose that we are conducting a study of academic performance in a Statistics course. Specifically suppose that we want to examine the relationship between grades in Statistics and grades in a Calculus course, grades in a Physics course and gender. So we collect the following information for each student in the study: student number, grade in Statistics, grade in Calculus, grade in Physics and

| | C1 | C2 | C3 | C4 | C5-T | C6 | C7 | C8 | C9 | ⋯ |
|---|---|---|---|---|---|---|---|---|---|---|
| | | | | | | | | | | |
| **1** | 12389 | 81 | 85 | 78 | m | | | | | |
| **2** | 97658 | 75 | 72 | 62 | m | | | | | |
| **3** | 53546 | 77 | 83 | 81 | f | | | | | |
| **4** | 55542 | 63 | 42 | 55 | m | | | | | |
| **5** | 11223 | 71 | 82 | 67 | f | | | | | |
| **6** | 77788 | 87 | 56 | * | f | | | | | |
| **7** | 44567 | 23 | 45 | 35 | m | | | | | |
| **8** | 32156 | 67 | 72 | 81 | m | | | | | |
| **9** | 33456 | 81 | 77 | 88 | f | | | | | |
| **10** | 67945 | 74 | 91 | 92 | f | | | | | |
| ⋮ | | | | | | | | | | |

Table 1: The `marks` worksheet

gender. Therefore we have 5 variables with student number and the grades in the three subjects being *numeric variables* and gender a *text variable*. Let us further suppose that there are 10 students in the study.

Table 1 gives a possible outcome from collecting the data in such a study. Column C1 contains the student number and note that this is a categorical variable even though it is a number. The student number primarily serves as an identifier so that we can check that the data has been entered correctly. This is something you should always do as a first step in your analysis. Columns C2-C4 contain the student grades in their Statistics, Calculus and Physics courses and column C5 contains the gender data. Notice that a column contains the values collected for a single variable and a row contains the values of all the variables for a single student. Sometimes a row is referred to as an *observation* or *case*. Observe that the data for this study occupies a $10 \times 5$ subtable of the full worksheet. All of the other blank entries of the worksheet can be ignored as they are undefined.

There will be limitations on the number of columns and rows you can have in your worksheet and this depends on the particular implementation of Minitab you are using. So if you plan to use Minitab for a large problem you should check with the system person or further documentation to see what these are. For example, in some versions of Minitab there is a limitation of 3500 cells. So there can be one variable with 3500 values in it or 35 variables with 100 values each, etc.

Associated with a worksheet is a table of *constants*. These are typically numbers that you want to use in some arithmetical operation applied to every value in a column. For example, you may have recorded heights of people in inches and want to convert these to heights in centimeters. Then you must multiply every height by the value 2.54. The Minitab constants are labeled K1, K2, etc. Again there are limitations on the number of constants you can associate with a worksheet. For example, in many versions there can be at most 1000 constants. So to continue with the above problem we might assign the value 2.54 to K1. In I.7.4 we show how to make such an assignment and in I.10.1 we show how to multiply every entry in a column by this value.

In Version 12 of Minitab there is an additional structure beyond the worksheet called the *project*. A project can have multiple worksheets associated with it. Also a project can have associated with it various graphs and records of the commands you have typed and the output obtained while working on the worksheets. Projects can be saved and retrieved for later work. Projects are discussed in Appendix A.

# 6 Minitab Commands

We will now begin to introduce various Minitab commands to get data into a worksheet, edit a worksheet, perform various operations on the elements of a worksheet and save and access a saved worksheet. Before we do, however, it is useful to know something about the basic structure of all Minitab commands. Associated with every command is of course its *name,* as in **exit** and **help**. Most commands also take *arguments* and these arguments are column names, constants and perhaps a file name enclosed in single quotes.

We first describe the use of commands in the session environment. If there is more than one argument then the arguments must be separated by commas or blanks. So the basic structure of such a command with $n$ arguments is

**command name** $E_1, E_2, ..., E_n$

where $E_i$ is the *ith* argument. Alternatively we can write

**command name** $E_1\ E_2\ ...\ E_n$

if we don't want to type commas. Conveniently if the arguments $E_1, E_2, ..., E_n$ are consecutive columns in the worksheet we have the following short-form

**command name** $E_1$-$E_n$

which saves even more typing and accordingly decreases our chance of making a typing mistake. If you are going to type a long list of arguments and you don't want them all on the same line then you can type the *continuation symbol* & where you want to break the line and hit Enter. Minitab responds with the prompt

CONT>

and you continue to type argument names. The command is executed when you hit Enter after an argument name without a continuation character following it.

Many commands can in addition be supplied with various subcommands which alter the behavior of the command. The structure for commands with subcommands is

**command name** $E_1$ ... $E_{n_1}$;
**subcommand name** $E_{n_1+1}$ ... $E_{n_2}$;
$\quad \vdots$
**subcommand name** $E_{n_{k-1}+1}$ ... $E_{n_k}$.

Notice that when there are subcommands each line ends with a semicolon until the last subcommand which ends with a period. Also subcommands may have arguments. When Minitab encounters a line ending in a semicolon it expects a subcommand on the next line and changes the prompt to

SUBC >

until it encounters a period whereupon it executes the command. If while typing in one of your subcommands you suddenly decide that you would rather not execute the subcommand – perhaps you realize something was wrong on a previous line – then type **abort** after the SUBC > prompt and hit Enter. As a further convenience it is worth noting that you only need to type in the first four letters of any Minitab command or subcommand.

In the window environment commands can be accessed as described above in the Session window or you can make use of the File, Edit, Manip, Calc, Stat, Graph and Editor entries in the menu bar. Clicking any of these brings up a list of commands that you can use to operate on your worksheet. The lists which appear may depend on which window is active, e.g. either a Data window or the Session window. Unless otherwise specified we will always assume that the Session window is active when discussing menu commands. If a command name in a list is faded then it is not available.

Typically using a command from the menu bar requires the use of *dialog boxes* which open when you click on a command in the list and which you use to provide the arguments and subcommands to the command and specify where the output is to go. Dialog boxes have various text boxes that must be filled in to correctly execute a command. Clicking on a box brings up a *variable list* on the left of all items in the active worksheet that can be placed in the text box. Double clicking on items in the variable list places them in the text box or you can type them in. When you have filled in the dialog box and clicked OK the command is printed in the Session window and executed. In general we refer readers to Help for more information on the use of commands from the menu bar, although we will always point out where each command can be found on the menu bar. Also dialog boxes have a Help button that can be used to learn how to make the entries. Quite often it is faster and more convenient to type your commands directly into the Session window as we have described above. If you do choose to use the menu commands rather than session commands you should still read the descriptions of the session commands in the manual as these provide sufficient information to fill in the dialog boxes appropriately.

In Version 12 there are two additional ways in which you can input commands to Minitab. Instead of typing the commands directly in to the Session window you can also type these directly into the Command Line Editor which is available via Edit ▶ Command Line Editor. Multiple commands can then be typed directly into a box that pops up and executed when the Submit Commands button is clicked. Output appears in the Session window. Also many commands are available on a *toolbar* that lies just below the menu bar at the top of the Minitab window. There is a different toolbar depending upon which window is active. We do not discuss the use of the toolbars in this manual.

# 7 Entering Data into a Worksheet

There are various methods for entering data into a worksheet. If you are in the window environment then you can use the *Data window* to enter data directly into the worksheet by clicking your mouse in a cell and then typing the corresponding data entry and hitting Enter. Remember that you can make a Data window active by clicking anywhere in the window or by using Windows in the menu bar. If you type any character that is not a number Minitab automatically identifies the column containing that cell as a text variable and indicates that by appending T (an A in older versions of Minitab) to the column name, e.g. C5-T in Table 1. You don't need to append the T when referring to the column. Also there is a *data direction arrow* in the upper left corner of the data window that indicates the direction the cursor moves after you hit Enter. Clicking on it alternates between row-wise and column-wise data entry. Certainly this is an easy way to enter data when it is available. Remember columns are variables and rows are observations! Also in Version 12 you can have multiple data windows open and move data between them. If you do not have the window environment available or your data is in an external file then you will need to learn about the **read** command.

# 7.1 The READ Command

In the session environment we use the **read** command to input data. The following is a use of the **read** command to create the first four columns in the worksheet in Table 1. Because C5 contains text data we have to treat it a little differently.

```
MTB >read c1-c4
DATA>12389 81 85 78
DATA>97658 75 72 62
DATA>53546 77 83 81
DATA>55542 63 42 55
DATA>11223 71 82 67
DATA>77788 87 56 *
DATA>44567 23 45 35
DATA>32156 67 72 81
DATA>33456 81 77 88
DATA>67945 74 91 92
DATA>end
 10 rows read.
```

Note that it doesn't matter if we use lower or upper case for the column names as Minitab is case insensitive. If you are inputting many variables you can use the continuation character & to continue inputting an observation (row) on the next line. A new observation begins after we hit Enter without a preceding &.

Also note that the sixth entry in C4 is *. This is the symbol Minitab uses to indicate a *missing numeric value*. A *missing text value* is simply denoted by a blank. This means that there was no data to record in this cell – perhaps the student dropped the course. Missing values are a common occurrence in real-world data sets and special attention should be paid to them. In general Minitab statistical analysis commands ignore any cases that contain missing data except that the output of the command will tell you how many cases were ignored because of missing data. It is important to pay attention to this information. If your data is riddled with a large number of missing values your analysis may be based on very few observations even if you have a large data set!

Also notice that we indicate to Minitab that we have no more data to input by typing **end** after the DATA> prompt and hitting Enter. Minitab then tells us how many rows we read in and returns us to the MTB > prompt and ready to type more commands.

We now consider entering the gender data in C5. Of course we could have recoded this using 1 to denote female and 0 to denote male, but sometimes this is not feasible. The following command reads in C5.

```
MTB >read c5;
SUBC>format (a1).
DATA>m
DATA>m
DATA>f
DATA>m
DATA>f
DATA>f
DATA>m
DATA>m
DATA>f
DATA>f
DATA>end
  10 rows read.
```

Notice that we have used the subcommand **format** with **read.** The (a1) that accompanies the format statement tells Minitab that we are going to read in an text variable that is 1 character long. Alternatively we could have read in the entire worksheet using the following Minitab command.

```
MTB >read c1-c5;
SUBC>format (f5.0,1x,f2.0,1x,f2.0,1x,f2.0,1x,a1).
DATA>12389 81 85 78 m
DATA>97658 75 72 62 m
    ⋮
```

Here the format statement says that we are going to type in the data according to the following rule: a numeric variable occupying 5 spaces and with no decimals, followed by a space, a numeric variable occupying 2 spaces with no decimals, a space, a numeric variable occupying 2 spaces with no decimals, a space, a numeric variable occupying 2 spaces with no decimals, a space, and a text variable occupying 1 space. This typing rule must be rigorously adhered to or errors will occur. So the rules you need to remember if you choose to use the **read** command together with the **format** subcommand are that ak indicates a text variable occupying k spaces, kx indicates k spaces and fk.1 indicates a numeric variable occupying k spaces of which l are to the right of the decimal point. Note if a data value does not fill up the full number of spaces allotted to it in the format statement it must be right justified in its field. Also if a decimal point is included in the number then this occupies one of the spaces allocated to the variable and similarly for a negative or plus sign. There are many other features to the **format** subcommand that we will not discuss here.

Typically we try to avoid the use of the **format** subcommand as it is somewhat cumbersome but sometimes we must use it. This necessity often occurs when the data comes to us in a computer file and we want to read it into Minitab. For example suppose that the data in Table 1 are stored as ASCII characters in a computer file called marks.txt in columns with single spaces between them. Then we can read this data into the worksheet via the following command.

```
MTB >read 'marks.txt' c1-c5;
SUBC>format(f5.0,1x,f2.0,1x,f2.0,1x,f2.0,1x,a1).
Entering data from file:  marks.txt
 10 rows read.
```

Again the data must be entered into the file marks.txt exactly as prescribed in the format statement or errors will occur. Also, depending on the version of Minitab we are using, we may have to provide the full pathname of the file between single quotes and you should check on this with your systems person. Obviously this is the most convenient way to make entries in a worksheet when there is a lot of data. When all the data in the file is numeric and variables are separated by spaces then we do not need the **format** subcommand and need only give the name of the file containing the data and the columns the data is to be read into. It is also possible to *import* data from other programs such as Lotus 1-2-3 and Microsoft Excel but we do not discuss this feature.

The general syntax of the **read** command is

**read** $E_1$ ... $E_m$

where $E_1$ is either a column to contain data or file pathname enclosed in single quotes, if we are inputting from a file, and $E_2$, ..., $E_m$ are columns to contain data.

In the window environment you can input the contents of columns from an external file by File ▶ Other Files ▶ Import Special Text Data, filling in the dialog box and specifying the file containing the data when prompted. Typically the file will have the ending `.txt` or `.dat`. In older versions the relevant menu command is File ▶ Import ASCII Data. We refer the reader to Help for more information on this.

## 7.2 The SET Command

The **set** command is used to input data into a single column. Of course **read** can also be used for this but **set** has some features that make it very useful. In particular it allows for the use of various shortcuts when the data we want to input is *patterned*. For example, suppose we want C6 to contain the 10 entries 1, 2, 3, 4, 5, 5, 4, 3, 2, 1. Then the command

```
MTB >set c6
DATA>1:5
DATA>5:1
DATA>end
```

does this. Note that for integers $m$ and $n$ the expression $m : n$ is equivalent to entering the data $m, m + 1, \ldots , n$ when $m < n$ and $m, m - 1, ..., n$ when $m > n$ and $m$ when $m = n$.

In fact more complicated patterns can be accommodated. The expression $m : n/d$, where $d > 0$, expands to a list as above but with the increment of $d$ or $-d$, whichever is relevant, replacing 1 or $-1$. If $m < n$ then $d$ is added to $m$ until the next addition would exceed $n$ and if $m > n$ then $d$ is subtracted from $m$ until the next subtraction would be lower than $n$. The expression $k(m : n/d)$ repeats $m : n/d$ for $k$ times while $(m : n/d)l$ repeats each element in $m : n/d$ for $l$ times. The expression $k(m : n/d)l$ repeats $(m : n/d)l$ for $k$ times.

Also we can add elements in the parentheses. For example, the command

```
MTB >set c6
DATA>(1:2/.5 4:3/.2)
DATA>end
```

creates the column with entries 1.0, 1.5, 2.0, 4.0, 3.8, 3.6, 3.4, 3.2, 3.0. The multiplicative factors $k$ and $l$ can also be used in such a context. Obviously there is a great deal of scope for entering patterned data with **set**.

We can use **set** to read in numeric data from a file into a single column just as we did with **read**. Also we can read text data into a single column with **set** by using the **format** subcommand as we did with **read**. The general syntax of the set command is

    **set** $E_1$

where $E_1$ is a column.

In the window environment the capability to enter patterned data can also be accessed via Çalc ▶ Make Patterned Data and includes the ability to enter patterned text data. In older versions the command is Çalc ▶ Set Patterned Data and text data are not allowed.

# 7.3 The PRINT Command

Once we have entered the data into the worksheet we should always check that we have made the entries correctly. Typically this means printing out the worksheet and checking the entries. The **print** command prints out the worksheet on the screen, or in the windows environment, in the Session window.

```
MTB >print c1-c5
 ROW    C1 C2 C3 C4 C5
```

```
 1 12389 81 85 78   m
 2 97658 75 72 62   m
 3 53546 77 83 81   f
 4 55542 63 42 55   m
 5 11223 71 82 67   f
 6 77788 87 56  *   f
 7 44567 23 45 35   m
 8 32156 67 72 81   m
 9 33456 81 77 88   f
10 67945 74 91 92   f
```

If the worksheet is quite extensive this may not be the most desirable way to check that your entries are correct as the worksheet will not fit on a single screen or window. Rather we would prefer to have the worksheet printed on paper so that we can do the checking more easily. We will discuss how to do this in I.9. The general syntax for the **print** command is

   **print** $E_1$ ... $E_m$

where $E_1$, ..., $E_m$ are columns and constants.

   In the window environment the menu command Manip ▶ Display Data will print the data you ask for in the dialog box to the Session window. In older versions this command is File ▶ Display Data.

# 7.4 The LET Command

To enter constants we must use the **let** command. For example, suppose that we wished to compute a weighted average of the grades in the `marks` worksheet for each student where we gave the Statistics mark a weight of .5 and the Calculus and Physics marks each a weight of .25. We will see how to compute this in the section on Arithmetical Operations on Variables and Constants but to do this we would like the constants K1, K2, K3 to equal .5, .25, .25 respectively. The following commands make this assignment and then we check, using the **print** command, that we have entered the constants correctly.

```
MTB >let k1=.5
MTB >let k2=.25
MTB >let k3=.25
MTB >print k1-k3
```

```
K1 0.500000
K2 0.250000
K3 0.250000
```

We will discuss other features of the let command in I.7.7 and I.10.1.

In the window environment this capability of the let command can be accessed using the menu command C̲alc ▶ Ca̲lculator and filling in the dialog box appropriately. Also we can assign constants text values. For example,

```
MTB >let k4=''result''
```

assigns K4 the value `result`. Note the use of double quotes. In older versions the command is C̲alc ▶ Mathematical E̲xpressions and then filling in the dialog box appropriately. We cannot assign constants text values in older versions.

# 7.5 The NAME Command

It often makes good sense to give the columns and constants names rather than just referring to them as C1, C2, ..., K1, K2 etc. This is especially true when there are many variables and constants as it would be easy to slip and use the wrong column in an analysis and then wind up making a mistake. To accomplish this we use the **name** command. For cxample in the `marks` worksheet,

```
MTB >name c1 'studid' c2 'stats' c3 'calculus' &
CONT>c4 'physics' c5 'gender' &
CONT>k1 'weight1' k2 'weight2' k3 'weight3'
```

gives the names `studid` to C1, `stats` to C2, `calculus` to C3, `physics` to C4, `gender` to C5, `weight1` to K1, `weight2` to K2 and `weight3` to K3. All of these names are evocative of their respective contents and so make it easy to remember which variable is which. Notice that we have made use of the continuation character & for convenience in typing in the full input to **name**. Variable and constant names can be at most 8 characters in length in many versions of Minitab, although this constraint is extended in Version 12 to 31 characters, cannot include the characters # and ' and cannot start with a leading blank or *. Recall that Minitab is not case sensitive so it does not matter if we use lower or upper case letters when specifying the names. When using the variables as arguments just enclose the names in single quotes. For example,

```
MTB >print 'studid' 'calculus'
```

prints out the contents of these variables on the screen or in the Session window.

In the window environment variable names can be entered directly on the Data window using the first space below the column designator (see Table 1) and the quotes are not required.

## 7.6 The INFO Command

We can get information on the data we have entered into the worksheet by using the **info** command. For example, we get the following results based on what we have entered into the **marks** worksheet so far.

```
MTB >info
  Column      Name        Count Missing
A C1          studid         10       0
  C2          stats          10       0
  C3          calculus       10       0
  C4          physics        10       1
A C5          gender         10       0
  Constant    Name        Value
  K1          weight1     0.500000
  K2          weight2     0.250000
  K3          weight3     0.250000
```

Notice that the **info** command tells us how many missing values there are and in what columns they occur and also the values of the constants.

In the window environment this information can be accessed directly from the *Info window* available via Window ▶ Info.

## 7.7 Editing a Worksheet

It often happens that after data entry we notice that we have made some mistakes or we obtain some additional information such as more observations. So far the only way we could change any entries in the worksheet or add some rows is to reenter the whole worksheet! Fortunately we don't have to do this as Minitab has a number of commands we can use.

The **restart** command is convenient if you should decide that you would like to start entering your data all over again or you have finished with the current worksheet and want to start a new one. The command

    MTB >restart

clears all entries, columns and constants, from the worksheet. Of course you can always do this by exiting Minitab and then access Minitab again but **restart** is more convenient. Remember, however, you should in general always save a worksheet you have been working on as this will save you from having to reenter all the data again. So always save the contents of the current worksheet before using **restart** unless you are absolutely sure you don't need to. We discuss how to save the contents of a worksheet in I.8.1.

In the window environment you can use File ▶ New to clear the worksheet. In older versions the command is File ▶ Restart Minitab.

Suppose, however, that we simply want to change an entry in a column. To do this we can use the **let** command. For example, if we want to change the Calculus mark for the student with **studid** 97658 from 72 to 85 in the **marks** worksheet – i.e. the second entry of the third row – the command

    MTB >let c3(2)=85

does this. Alternatively we could have used

    MTB >let 'calculus'(2)=85

as a safer method as this ensures we don't get the columns mixed up and so we change the right variable. In general ci(k) refers to the *kth* entry in the *ith* column. Note that the **let** command cannot be used in this way to edit text columns in many versions of Minitab but can in Version 12. This is another reason to try to avoid text data in a worksheet or to convert it to numeric, when there is no option, using the **convert** command covered in I.11.3.

While the **let** command allows you to carry out very simple editing chores large scale editing tasks require more powerful commands. A number of such commands are covered in I.11. The **insert** command, covered in I.11.7, allows you to insert rows or observations anywhere in the worksheet and the **delete** command, covered in I.11.5, allows you to delete rows and the **erase** command, covered in I.11.6, is for the deletion of columns or variables from the worksheet.

In the window environment editing the worksheet is straightforward since we simply make the Data window active and change any cells by retyping

their entries and hitting the Enter key. We can add rows and columns at
the end of the worksheet by simply typing new data entries in the relevant
cells. See the sections on the **insert**, **delete** and **erase** commands for more
discussion on other editing features available in this environment.

# 8 Saving and Retrieving a Worksheet

Quite often you will want to save the results of all your work in creating the
worksheet. If you exit Minitab before you save your work  you will have to
reenter everything. So we recommend that you always save.

## 8.1 The SAVE Command

To do this you give the worksheet a file name and use the **save** command.
Recall the restrictions on file names discussed in I.3. For example, suppose
that we want to save the worksheet we created in I.7.1 and give it the name
`marks`. The following command does this.

```
MTB >save 'marks'
Saving worksheet in file:  marks.MTW
```

Notice that the file saved on  the hard drive is called `marks.mtw`. The extension `.mtw` signifies that this is a Minitab worksheet. This is a formatted file
and you will not want to edit or try to print this file outside of Minitab. We
discuss how to print a hard copy of a worksheet in I.9. If you give the **save**
command without a file name then the worksheet is saved using the default
name `minitab`.

In the window environment to save a worksheet use File ▶ Save Current
Worksheet to save the worksheet with its current name or the default name
if it doesn't have one and File ▶ Save Current Worksheet As if you wish to
provide a name.

## 8.2 The RETRIEVE Command

If you start a new session and want to work again on a worksheet that you
have saved you can use the **retrieve** command. For example,

```
MTB >retrieve 'marks'
```

```
Retrieving worksheet from file:  marks.MTW
Worksheet was saved on 2/ 7/1998
```

establishes the worksheet saved as **marks** as the current active worksheet
in Minitab; all the columns and constants that you previously entered are
just as they were before. Minitab typically comes with a number of example worksheets that can be used for problems and examples. These can be
accessed using the **retrieve** command. See your instructor, systems person
or documentation to see what worksheets come with the version of Minitab
you are using.

In the window environment the menu command File ▶ Open Worksheet
leads to a list of available worksheets one of which you can select by clicking
on its name.

# 9 Recording a Session and Printing Worksheets

Sometimes it is useful – e.g. when you have to hand in an assignment – to
maintain a record of all the commands you typed, the output you obtained
and any comments you want to make on what you are doing in a Minitab
session. If you want to insert a *comment* in the Session window then type
# after the MTB > prompt and then type your comment. Minitab is ready
to accept a command again as soon as you hit enter. Of course you could
place a comment on the new line as well.

In Version 12 you can edit the worksheet after you have completed your
work using the menu command Editor ▶ Make Output Editable or turn this
feature off via the command Editor ▶ Make Output Read-Only.

## 9.1 The OUTFILE Command

If we want to have access to everything that occurred in a session – e.g. you
want to print it out to hand in – then use the **outfile** command together
with a file name. For example,

```
MTB >outfile 'assign1'
Collecting Minitab session in file:  assign1.LIS
MTB ># Assignment 1
 ⋮
```

```
MTB >nooutfile
```

records all the commands, output and comments that occur in the session
in a file called `assign1.lis` until you type the command **nooutfile**. Note
the single quotes around the file name. The first line in this file is a comment.

```
MTB ># Assignment 1
```

The directory where `assign1.lis` is stored depends upon the particular
implementation of Minitab that you are using and you will have to check
with your documentation, instructor or systems person to determine this.
You can then print out `assign1.lis` as an ASCII text file using a method
relevant to the system you are using.

In the window environment there are several alternative methods you can
use to record a session. For example, if you don't want a permanent record of
the session you can use the command File ▶ Print Session Window. There
are some limitations on how much is retained so don't use this method if
you are planning an extensive session. To actually record the session in a
file use File ▶ Save Session Window As. In older versions the command File
▶ Other Files ▶ Start Recording Session works like **outfile**. We refer the
reader to the Help documentation for further details.

Therefore using the **outfile** command gives one method for printing out
the contents of a worksheet. For after giving this command we can use the
**print** command to print the worksheet to the screen or session window and
then use the system printer to print the output file. The output could be
quite voluminous, however, depending on the amount of data you have. If
you have a lot of data it might be better to use the **write** command discussed
in I.9.2.

In the window environment the worksheet can be printed by making the
Data window active and then using File ▶ Print Session Window. Again if
the data set is large it may be better to write the data out to a file. This
can be accomplished using the methods discussed in I.9.2.

## 9.2 The WRITE Command

Sometimes it is desirable to print the contents of a worksheet to an external
computer file as ASCII text. For this we use the **write** command. For
example, the command

```
MTB >write 'out' c1-c5
```

```
Writing data to file:  out.DAT
```

writes the contents of the worksheet to a file called `out.dat` and this file can be edited using your usual editor and printed on your system printer. For example, the contents of the file `out.dat,` when `marks` is the worksheet and we have performed the above command, are

```
12389 81 85 78 m
97658 75 72 62 m
53546 77 83 81 f
55542 63 42 55 m
11223 71 82 67 f
77788 87 56  * f
44567 23 45 35 m
32156 67 72 81 m
33456 81 77 88 f
67945 74 91 92 f
```

and note that there are no column or row labels. If you want the data to appear differently in the file then you can use the **format** subcommand to specify how you want the data written out.

In the window environment you can output the contents of columns to an external file using File ▶ Other Files ▶ Export Special Text, filling in the dialog box and specifying the destination file when prompted. In older versions this is accomplished by File ▶ Export ASCII Data, filling in the dialog box and specifying the destination file when prompted.

# 10 Mathematical Operations

When carrying out a data analysis a statistician is often called upon to transform the data in some way. This may involve applying some simple transformation to a variable to create a new variable – e.g. take the natural logarithm of every grade in the `marks` worksheet – to combining several variables together to form a new variable – e.g. calculate the average grade for each student in the `marks` worksheet. In this section we present some of the ways of doing this.

# 10.1 Arithmetical Operations

Simple arithmetic can be carried out on the columns of a worksheet using the arithmetical operations of addition +, subtraction −, multiplication *, division / and exponentiation ** together with the **let** command. When columns are added together, subtracted one from the other, multiplied together, divided one by the other (make sure there are no zeros in the denominator column) or one column exponentiates another, these operations are always performed component-wise. For example, C1*C2 means that the *ith* entry of C1 is multiplied by the *ith* entry of C2; etc. Also make sure that the columns you are going to perform these operations on correspond to numeric variables! While these operations have the order of precedence **, */, +−, parentheses ( ) can and should be used to ensure an unambiguous result. For example, suppose in the **marks** worksheet we wanted to create a new variable by taking the average of the Statistics and Calculus grades and then subtracting this from the Physics grade and then place the result in C6. Then the command

```
MTB >let c6=c4-(c2+c3)/2
```

accomplishes this. Of course we could have substituted the variable names, in single quotes, for C2, C3 and C4. These operations are done on the corresponding entries in each column; corresponding entries in the columns are operated on according to the formula we have specified and a new column of the same length containing all the outcomes is created. Note that the sixth entry in C6 will be * − missing − because this entry was missing for C4.

We can also use these arithmetical operation on the constants K1, K2, etc. and numbers to create new constants or use the constants as *scalars* in operations with columns. For example, suppose that we want to compute the weighted average of the Statistics, Calculus and Physics grades where Statistics gets twice the weight of the other grades. Recall that we created, as part of the **marks** worksheet the constants **weight1** = .5, **weight2** = .25 and **weight3** = .25 in K1, K2 and K3, respectively. So this weighted average is computed via the command

```
MTB >let c7='weight1'*'stats'+'weight2'*'calculus'&
CONT>+'weight3'*'physics'
```

and the result is placed in C7. We have used the continuation character & for convenience in this computation.

The arithmetical operations can also be carried out using the **add, subtract, multiply, divide** and **raise** commands. For example, if $E_1$, $E_2$ and

$E_3$ are columns of the same dimension then

> **add** $E_1$ $E_2$ $E_3$

adds $E_1$ row-wise to $E_2$ and places the result in $E_3$; i.e. each element of $E_1$ is added to the corresponding element of $E_2$ and the result is placed in the corresponding entry of $E_3$. Similarly the commands,

> **subtract** $E_1$ $E_2$ $E_3$
> **multiply** $E_1$ $E_2$ $E_3$
> **divide** $E_1$ $E_2$ $E_3$
> **raise** $E_1$ $E_2$ $E_3$

respectively subtract $E_2$ row-wise from $E_1$, multiply $E_1$ row-wise by $E_2$, divide $E_1$ row-wise by $E_2$ raise $E_1$ row-wise by $E_2$ and place the results in $E_3$. If the operation is impossible then the result is set to *, missing. If either $E_1$ or $E_2$ is constant then the constant is applied to each element; for example,

```
MTB > raise c1 2 c2
```

squares each element of C1 and places the result in C2 while

```
MTB > raise 2 c1 c3
```

raises 2 to the powers given in C2 and places the result in C3. The **add** and **multiply** commands operate slightly more generally than we have described and we refer the reader to **help** for further details.

In the window environment these computations can also be carried out using Calc ▶ Calculator and filling in the dialog box appropriately. In older versions of Minitab the command is Calc ▶ Mathematical Expressions.

# 10.2 Mathematical Functions

Various mathematical functions are available in Minitab. For example, suppose we want to compute the natural logarithm of the Statistics mark for each student. Then the command

```
MTB >let c8=loge(c2)
```

does this and places the result in C8. There are a number of such functions and a complete list is provided in Appendix B.1. These functions can be applied to numbers as well as constants. If you want to know the sine of the number 3.4 then

```
MTB >let k4=sin(3.4)
MTB >print k4
K4 -0.255541
```

gives the value.

In the window environment these computations can also be carried out using Çalc ▶ Calculator and filling in the dialog box appropriately. In older versions the command is Çalc ▶ Mathematical Expressions.

## 10.3 Column and Row Statistics

There are various *column statistics* which compute a single number from a column by operating on all of the elements in a column. For example, suppose that we want the mean of all the Statistics marks, i.e. the mean of all the entries in C2. Then the command

```
MTB >mean(c2)
 MEAN = 69.900
```

gives this value. Note that for commands that compute column statistics we do not need to store the value in a constant and then print it out. On the other hand if we want to save the value for subsequent use then the command

```
MTB >let k1=mean(c2)
```

does this. The general syntax for column statistic commands is

**column statistic name**$(E_1)$

where the operation is carried out on the entries in column $E_1$ and output is written to the screen unless it is assigned to a constant using the **let** command. See Appendix B.2 for a list of all the column statistics available.

Also for most column statistics there are versions that compute *row statistics* and these are obtained by placing **r** in front of the column statistic name. For example,

```
MTB >rmean(c2 c3 c4 c6)
```

computes the mean of the corresponding entries in C2, C3 and C4 and places the result in C6. The general syntax for row statistic commands is

**row statistic name**$(E_1 \ldots E_m \ E_{m+1})$

where the operations are carried out on the rows in columns $E_1, \ldots, E_m$ and the output is placed in column $E_{m+1}$. See Appendix B.3 for a list of all the row statistics available.

In the window environment these computations can also be carried out using Çalc ▶ Calculator or using Çalc ▶ Column Statistics and Çalc ▶ Row Statistics respectively and filling in the dialog box appropriately.

# 10.4 Comparisons and Logical Operations

Minitab also contains the following comparison and logical operators.

| Comparison Operators | Logical Operators |
|---|---|
| equal to =, **eq** | &, **and** |
| not equal to <>, **ne** | \, **or** |
| less than <, **lt** | ˜, **not** |
| greater than >, **gt** | |
| less than or equal to <=, **le** | |
| greater than or equal to >=, **ge** | |

Notice that there are two choices for these operators; for example, either use the symbol >= or the mnemonic **ge**.

The comparison and logical operators are useful when we have simple questions about the worksheet which would be tedious to answer by inspection. This feature is particularly useful when we are dealing with large data sets. For example, suppose that we want to count the number of times the Statistics grade was greater than the corresponding Calculus grade in the **marks** worksheet. Then

```
MTB >let c6=c2>c3
MTB >let k4=sum(c6)
MTB >print k4
K4 4.00000
```

accomplishes this. How does it work? First note that the comparisons operate by comparing each element in a column with its corresponding element in the other column. So in the above command the *ith* entry in C2 is compared to the *ith* entry in C3 and if it is greater a 1 is recorded in the *ith* place in C6 and a 0 is recorded there otherwise. Therefore a 1 is recorded whenever the condition is *true* and a 0 is recorded whenever the condition is *false*. The

**sum** function adds up all the entries in a column so in this case, because each entry in C6 is a 1 or 0, the sum just records the number of times the condition was satisfied.

The logical operators combine with the comparison operators to allow more complicated questions to be asked. For example, suppose we wanted to calculate the number of students whose Statistics mark was greater than their Calculus mark and less than or equal to their Physics mark. The commands

```
MTB >let c6=c2>c3 and c2<=c4
MTB >let k4=sum(c6)
MTB >print k4
K4 1.00000
```

accomplish this. In this case both conditions c2>c3 and c2<=c4 have to be true for a 1 to be recorded in C6. Note that the observation with the missing Physics mark is excluded.

In the window environment these computations can also be carried out by clicking on Calc ▶ Calculator and filling in the dialog box appropriately. In older versions the command is Calc ▶ Mathematical Expressions.

In Version 12 text variables can be used in comparisons where the ordering is alphabetical. For example,

```
MTB >let c6=c5<''m''
```

puts a 1 in C6 whenever the corresponding entry in C5 is alphabetically smaller than m.

# 11 Some More Minitab Commands

In this section we discuss some commands that can be very helpful in certain applications. We will make reference to these commands at appropriate places throughout the manual and it is probably best to wait to read these descriptions until such a context arises.

## 11.1 The CODE Command

The **code** command is used to recode numeric columns. By this we mean that data entries in columns are replaced by new values according to a coding scheme that we must specify. For example, suppose in the **marks** worksheet we want to recode the grades in C2, C3 and C4 so that any mark in the range 0-9 becomes a 0, every mark in the range 10-19 becomes 10, etc. and the results are placed in columns C6, C7 and C8. The following command

```
MTB >code(0:9) to 0 (10:19) to 10 (20:29) to 20 (30:39) to 30 &
CONT>(40:49) to 40 (50:59) to 50 (60:69) to 60 (70:79) to 70 &
CONT>(80:89) to 80 (90:99) to 90 for C2-C4 put in C6-C8
```

accomplishes this. Note the use of the continuation symbol & as this is a long command. The notation $d : e$ denotes the inclusive range from $d$ to $e$. Suppose now we wish to recode the data in C6-C8 so that 0 and 10 become 5, 20 and 30 become 25, etc. and we place the result back in C6-C8. Then the command

```
MTB >code (0,10) to 5 (20,30) to 25 (40,50) to 45 (60,70) to 65&
CONT>(80,90) to 85 for c6-c8 put in c6-c8
```

accomplishes this. The general syntax for the **code** command is

**code** $(V_1)$ to $code_1$ ... $(V_n)$ to $code_n$ for $E_1$ ... $E_m$ put in $E_{m+1}$ ... $E_{2m}$

where $V_i$ denotes a set of possible values and ranges for the values in columns $E_1$ ... $E_m$ that are all coded as the number $code_i$ and the results of this coding are placed in the columns $E_{m+1}$ ... $E_{2m}$ ; i.e., the recoded $E_1$ is placed in $E_{m+1}$ etc. Note that if a value in a column is not covered by one of the sets $values_i$ then it is simply left the same in the new column. The **code** command can also be used to recode missing values by including * in one of the sets $V_i$.

In the window environment recoding can also be carried out using Manip ▶ Code and note that any recoding can be made – i.e. numeric to numeric, numeric to text, etc. In older versions the command is Manip ▶ Code Data Values and it is restricted to numeric to numeric recodings as is the **code** session command.

# 11.2 The CONCATENATE Command

The **concatenate** command combines two or more text columns into a single text column. For example if C6 contains m, m, m, f, f, reading first to last entry, and C7 contains to, ta, ti, to, ta then the command

```
MTB >concatenate c6 c7 in c8
```

places the entries mto, mta, mti, fto, fta in C8. The general syntax of the **concatenate** command is

**concatenate** $E_1$ ... $E_m$ in $E_{m+1}$

where $E_1$, ..., $E_m$ are text columns and $E_{m+1}$ is the target text column. In some versions of Minitab there are restrictions on the length of text variables so there is a limit on the amount of concatenation we can do.

In the window environment concatenation can be accomplished using Manip ▶ Concatenate and filling in the dialog box appropriately.

## 11.3 The CONVERT Command

The **convert** command is used to change text data into numeric data and vice versa. As dealing with text data is a bit more difficult in Minitab, we recommend either converting text data to numeric before input or using the **convert** command after input to carry this out. This comment is much less relevant with Version 12.

For example, in the worksheet `marks` suppose we want to change the gender variable from text, with male and female denoted by m and f respectively, to a numerical variable with male denoted by 0 and female by 1. To do this we must first set up a *conversion table*. The conversion table comprises two columns in the worksheet where one column is text and contains the text values used in the text column and the second column is numeric and contains the numerical values that you want these changed into. For example, the commands

```
MTB >read c6 c7;
SUBC>format (a1,1x,f1.0).
DATA>m 0
DATA>f 1
DATA>end
 2 rows read.
MTB >convert c6 c7 c5 c8
MTB >print c5 c8
```

create a new text column C6 with the entries m, f and new numeric columns C7, with entries 0,1, and C8, with a 1 for every f in C5 and 0 for every m in C5. So the columns C6, C7 contain the conversion table and C8 contains the converted variable. Note that we could have converted C5 by replacing C8 with C5 in the command.

The general syntax for the **convert** command is

**convert** $E_1$ $E_2$ $E_3$ $E_4$

where $E_1$, $E_2$ are the columns containing the conversion table, $E_3$ is the column to be converted and $E_4$ is the column containing the converted column.

In the window environment the conversion can be carried out using Manip ▶ Code ▶ Use Conversion Table and filling in the dialog box appropriately. You must still have set up a conversion table first to use this command. In older versions the command is Manip ▶ Convert.

# 11.4 The COPY Command

The **copy** command is used to copy columns to columns, constants to constants, columns into constants, constants into columns. For example, suppose we want a copy of the Statistics grades in C2 placed in C6. Then the command

```
MTB >copy c2 into c6
```

does this.

The syntax of the **copy** command is given by

**copy** $E_1$ ... $E_m$ into $E_{m+1}$ ... $E_{2m}$

to copy columns (constants) $E_1$, ..., $E_m$ into columns (constants) $E_{m+1}$, ..., $E_{2m}$,

**copy** $E_1$ ... $E_m$ into $E_{m+1}$

to copy constants $E_1$, ..., $E_m$ into column $E_{m+1}$ and

**copy** $E_1$ into $E_2$ ... $E_m$

to copy column $E_1$ into constants $E_2$, ..., $E_m$.

The **copy** command together with subcommands can also be used to select a subset of the worksheet. For example, suppose in the `marks` worksheet we want the marks of all the female students in a separate set of columns. Then the **use** subcommand

```
MTB >copy c1-c5 into c6-c10;
SUBC>use 3 5 6 9 10.
```

copies the contents of rows 3, 5, 6, 9 and 10 into columns C6-C10 and we see from Table 1 that these are just the rows corresponding to the female students. Also the **omit** subcommand

```
MTB >copy c1-c5 into c6-c10;
SUBC>omit 1 2 4 7 8.
```

accomplishes this by copying all the rows into C6-C10 except for rows 1, 2, 4, 7 and 8 which contain the grades of the males.

An even easier way to select a subset of rows, as opposed to listing all the rows we want or don't want, is with the **use** or **omit** subcommands selecting rows based on the values in a certain column. In this case we would like to select based on the f values in C5 but unfortunately some versions of Minitab won't allow us to select using text columns. So for those versions we must first use the **convert** command to recode f and m as described in I.11.3. Then the commands

```
MTB >read c6 c7;
SUBC>format(a1,1x,f1.0).
DATA>m 0
DATA>f 1
DATA>end
 2 rows read.
MTB >convert c6 c7 c5 c6
MTB >copy c1-c5 into c6-c10;
SUBC>use c6=1.
```

first recode C5 into C6 with f becoming 1 and m becoming 0, then copies the contents of C1-C5 into C6-C10 but only those rows where the entry in C6 is 1, i.e. only those rows that correspond to female students. Note that we overwrite C6 but not before its values are used to select the rows to copy. The **omit** subcommand can also be used in this way. You can also use the missing value symbol * in **use** or **omit** provided you enclose it in single quotes as in

```
SUBC>omit c4='*'.
```

The general syntax for the **use** and **omit** subcommands is

> **use** $E_1 = E_2... E_m$
> **omit** $E_1 = E_2... E_m$

where $E_1$ is a column and $E_2$, ..., $E_m$ are numbers or constants used to select the rows. If you are using Version 12 then you can select using text columns simply by enclosing the values in the column you want to select on in double quotes, as in

```
SUBC>use c5=''f''.
```

In the window environment these copying operations can also be carried out using Manip ▶ Copy Columns and filling in the dialog box appropriately. Copying can also be done very efficiently in the Data window using various cutting and pasting operations and commands under Edit on the menu bar.

## 11.5 The DELETE Command

The **delete** command is used to remove rows or cells within rows. For example suppose we now want to delete the third row of the worksheet `marks`. Then the command

```
MTB >delete 3 of c1-c5
```

accomplishes this. The general syntax of the **delete** command is

$$\textbf{delete } R_1\ R_2\ ...\ R_m\ \text{of}\ E_1\ ...\ E_n$$

where $R_1$, ..., $R_m$ denote the rows and $E_1$, ..., $E_n$ denote the columns where the deletions are to take place. Notice, however, whenever we delete a cell the contents of the cells beneath the deleted one in that column simply move up to fill the cell. The cell entry does not become missing; rather, cells at the bottom of the column become undefined! If you delete an entire row then this is not a problem because the rows below just shift up. For example, if we delete the third row then in the new worksheet, after the **delete** command, the third row is now occupied by what was formerly the fourth row. Therefore you should be very careful in using the **delete** command, when you are not deleting whole rows, to ensure that you get the result you intended.

In the windows environment the deletion of rows can be accomplished by Manip ▶ Delete Rows and filling in the dialog box. Deleting can also be done very efficiently in the Data window using various cutting and pasting operations and commands under Edit on the menu bar.

## 11.6 The ERASE Command

The **erase** command is used to remove variables from the worksheet, a good idea if you have lots of variables and don't need some anymore. For example, to delete the Physics grades – C4 – from the `marks` worksheet the command

```
MTB >erase c4
```

does this and C4 is no longer in the worksheet. Note, however, that C5 still contains the values of the variable gender. Also note that even if you delete all the values from a variable, say using the **delete** command, the variable remains in the worksheet. The general syntax of the **erase** command is

   **erase** $E_1 \ldots E_n$

where $E_1 \ldots E_n$ is any selection of variables or constants.

   In the window environment the deletion of variables and constants can be accomplished from Manip ▶ Erase Variables and filling in the dialog box.

# 11.7 The INSERT Command

Suppose that we want to add observations to the worksheet. For this we use the **insert** command. For example, suppose another student's marks are to be added to the **marks** worksheet and for some reason we want this observation to appear as the third row. Then

```
MTB >insert 2 3 c1-c5;
SUBC>format(a5,1x,f2.0,1x,f2.0,1x,f2.0,1x,a1).
DATA>87345 63 66 71 m
DATA>end
  1 rows read.
```

adds a row with **studid** = 87345, **stats** = 63, **calculus** = 66, **physics** = 71 and **gender** = m. Minitab will continue to add observations until we indicate that we are no longer adding rows by typing **end** at the **DATA>** prompt. Notice that we needed the **format** subcommand here because some of the data, namely **c5**, is text. If all the data is numeric then we can dispense with this.

   The general syntax for the **insert** command is

   **insert** $R_1$ $R_2$ $E_1 \ldots E_n$

where $R_1$ is the row number after which we want the observations inserted, $R_2$ is the following row and $E_1, \ldots, E_n$ are the columns where we want the data entries to be made – we don't have to make entries in every column. If we don't make entries in every column then note that the data at and below row R2 in the columns we do make entries in gets shifted down. So in this case you have to be careful to make sure the command is doing what you want. If we want to insert observations at the start of the worksheet then take $R_1 = 0$ and $R_2 = 1$. If we want to add observations at the end of the worksheet then simply drop $R_1$ and $R_2$:

**insert** $E_1$ ... $E_n$

will add rows to the end of the worksheet.

In the window environment row insertions can be accomplished when the data window is active by placing the cursor in the row before which you want to add observation, accessing the command Editor ▶ Insert Rows and then making data entries to the relevant cells.

# 11.8 The JOURNAL Command

If you just want to keep a *history* of what commands you have used and what data you have entered in your session then you can use the **journal** command. For example,

```
MTB >journal 'comm1'
Collecting keyboard input(commands and data)in file:
                                                comm1.MTJ

MTB >read c1 c2 c3
DATA>1 2 3
DATA>end
 1 rows read.
MTB >nojournal
```

puts

```
read c1 c2 c3
1 2 3
end
nojournal
```

into the file comm1.mtj. The history is turned off as soon as the **nojournal** command is typed. Typically the **journal** command is used when you want to create a file of Minitab commands to be used in *macros* or *execs* which are discussed in Appendix C.

In the window environment the *History window* keeps such a record with some limitations on the amount of history retained. Entries in the history window can be cut and pasted to the Session window and edited before being executed again so that a number of commands can be executed at once without retyping. This is very helpful when you have typed a long sequence of commands and then realize you made an error early on. See the Help documentation for further details.

## 11.9 The RANK Command

Sometimes we want to compute the *ranks* of the numeric values in a column. The rank $r_i$ of the *ith* value in a column is a value that reflects its relative size in the column. For example, if the *ith* value is the smallest value then $r_i = 1$, if it is the third smallest then $r_i = 3$, etc. If values are the same, i.e. *tied*, then each value receives the average rank. To calculate the ranks of the entries in a column we use the **rank** command. For example, suppose that C6 contains the values 6, 4 , 3, 2, 3, 3, 1. Then the **rank** command

```
MTB >rank c6 c7
MTB >print c6 c7
 ROW C6    C7
   1  6   6.0
   2  4   5.0
   3  3   3.5
   4  2   2.0
   5  3   3.5
   6  1   1.0
```

calculates the ranks of the elements of C6 and places these ranks in C7. Notice that the ranks assigned to the third and fifth items are equal to the average rank $(3+4)/2=3.5$ as the values in C6 are the same.

The general syntax of the rank command is

**rank** $E_1$ $E_2$

where $E_1$ is the column whose ranks we want to compute and $E_2$ is the column that will hold the computed ranks.

In the window environment the ranks can be calculated using Manip ▶ Rank and then filling in a dialog box appropriately.

## 11.10 The SORT Command

It often occurs as part of a data analysis that we want to sort a column so that its values ascend from smallest to largest or descend from largest to smallest.

Note that ordering here could refer to numerical order or alphabetical order so we also consider ordering text columns. Also we may want to sort all the rows contained in some subset of the columns in the worksheet by a particular column. The **sort** command allows us to carry out these tasks.

For example, suppose that we want to sort the entries in C2 in the `marks` worksheet – the Statistics grades – from smallest to largest, and place the sorted values in C6. Then the command

```
MTB >sort c2 c6
```

accomplishes this. If instead we wanted these marks to be sorted from largest to smallest then using the subcommand **descending**

```
MTB >sort c2 c6;
SUBC>descending c2.
```

sorts the entries in C2 in descending order and places the result in C6. Alternatively suppose that we want to sort all the grades by the student number. Then

```
MTB >sort c1 carry c2-c4 into c6-c9
```

does this. We can also sort by several columns using the **by** subcommand. For example,

```
MTB >sort c1-c5 c6-c10;
SUBC>by c5 c1.
```

sorts the entire `marks` worksheet first by gender and then within gender by student number. Therefore the marks of females appear first in C6-C10, ordered by student number, followed by the marks of males again ordered by student number.

The general syntax of the **sort** command is

$$\text{\textbf{sort} } E_1 \text{ carry } E_2 \ldots E_m \text{ into } E_{m+1} \ldots$$

where $E_1$ is the column to be sorted and $E_2, \ldots, E_m$ are carried along with the results placed in columns $E_{m+1} \ldots$. Note that this sort can also be accomplished using the **by** subcommand where the general syntax is

$$\text{\textbf{sort} } E_1 \, E_2 \ldots E_m \text{ into } E_{m+1} \ldots E_{2m};$$
$$\text{\textbf{by} } E_{2m+1} \ldots E_n.$$

where now we sort by columns $E_{2m+1}$, ..., $E_n$, sorting first by $E_{2m+1}$, then $E_{2m+2}$, etc., carrying along $E_1, \ldots, E_m$ and placing the result in $E_{m+1}, \ldots, E_{2m}$.

The **descending** subcommand can also be used to indicate which sorting variables we want to use in descending order rather than ascending order.

In the window environment sorting can be carried by clicking on Manip ▶ Sort and filling in the dialog box appropriately.

## 11.11 The STACK Command

The **stack** command is used to literally stack columns one on top of the other. For example, in the **marks** worksheet the command

```
MTB >stack c2 c3 c4 into c6;
SUBC>subscripts c7.
```

constructs the column C6 by first putting the entries in C2 followed by C3 and then by C4, so that C6 has 30 entries, and in column C7 puts the value 1 for all those entries from C2, 2 for all those entries from C3 and 3 for all those entries from C4. The subcommand **subscripts** can be deleted if we have no need of these values.

The general syntax for the **stack** command is given by

$$\text{stack } E_1 E_2 \ldots E_m \text{ into } E_{m+1}$$

where $E_1$, $E_2$, ..., $E_m$ denote the columns or constants to be stacked one on top of the other, starting with $E_1$, and with the result placed in column $E_{m+1}$.

In the window environment stacking can be accomplished using Manip ▶ Stack/Unstack ▶ Stack Columns and filling in the dialog box appropriately. Note that in Version 12 it is also possible to stack rows.

It is also possible to simultaneously stack blocks of columns and we refer the reader to **help** or Help for information on this.

## 11.12 The UNSTACK Command

The **unstack** command is the reverse of the **stack** command: a single column is broken up into a number of columns. For example, in the **marks** worksheet if we have stacked C2, C3 and C4 into C6 with subscripts in C7, then

```
MTB >unstack c6 c8-c10;
SUBC>subscripts c7.
```

unstacks these columns into C8-C10. Note that the **subscripts** subcommand is always necessary with the **unstack** command as this shows where the information is on how to break the single column into separate columns.

The general syntax for the **unstack** command is

**unstack** $E_1$ into $E_2 \ldots E_m$;
**subscripts** $E_{m+1}$.

where $E_1$ is the column to be unstacked, $E_2$, ..., $E_m$ are the columns and constants to contain the unstacked column and $E_{m+1}$ gives the subscripts 1, 2, ... that indicate how $E_1$ is to be unstacked.

In the window environment unstacking can be accomplished using Manip ▶ Stack/Unstack ▶ Unstack One Column and filling in the dialog box appropriately.

Note that it is also possible to simultaneously unstack blocks of columns and we refer the reader to **help** or Help for information on this.

# 12 Exercises

1. The following data give the Hi and Low trading prices in Canadian dollars for various stocks on a given day on the Toronto Stock Exchange. Create a worksheet, giving the columns the same variable names, using any of the methods discussed in I.7. Be careful to ensure that the value of the variable **stock** starts with a letter. Print the worksheet to check that you have successfully entered it. Save the worksheet giving it the name **stocks**.

   | Stock | Hi | Low |
   |-------|--------|--------|
   | ACR | 7.95 | 7.80 |
   | MGI | 4.75 | 4.00 |
   | BLD | 112.25 | 109.75 |
   | CFP | 9.65 | 9.25 |
   | MAL | 8.25 | 8.10 |
   | CM | 45.90 | 45.30 |
   | AZC | 1.99 | 1.93 |
   | CMW | 20.00 | 19.00 |
   | AMZ | 2.70 | 2.30 |
   | GAC | 52.00 | 50.25 |

2. Retrieve the worksheet **stocks** created in Exercise 1. Change the Lo value in the stock MGI to 3.95. Calculate the average of the Hi and Low

prices for all the stocks and save this in a column called `average`. Calculate the average of all the Hi prices and save this in a constant called `avhi`. Similarly do this for all the Lo prices and save this in a constant called `avlo`. Save the worksheet using the same name. Write all the columns out to a file called `stocks.dat`. Print the file `stocks.dat` on your system printer.

3. Retrieve the worksheet created in Exercise 2. Using the Minitab commands discussed in I.10 calculate the number of stocks in the worksheet whose `average` is greater than $5.00 and less than or equal to $45.00.

4. Using the worksheet created in Exercise 2 use the **insert** command to add the following stocks to the beginning of the worksheet.

| Stock | Hi | Lo |
|-------|-------|-------|
| CLV | 1.85 | 1.78 |
| SIL | 34.00 | 34.00 |
| AC | 14.45 | 14.05 |

Using the **erase** command delete the variable `average`. Save the worksheet.

5. Using the worksheet created in Exercise 4 and the **sort** command, sort the stocks into alphabetical order. Using the **rank** command calculate the ranks of the individual stocks based on their Hi price and save the ranking in a new column. Save the worksheet.

6. Using the worksheet created in Exercise 5 and the **copy** command calculate the average Hi price of all the stocks beginning in A. Recall that the **copy** command may not be able to be used to select a subset using a text column depending on the version of Minitab you are using.

7. Using the worksheet created in Exercise 5 and the **code** command recode all the Lo prices in the range $0-9.99 as 1, in $10-39.99 as 2 and greater than or equal to $40 as 3 and save the recoded variable in a new column.

8. Using the **set** command place the values from $-10$ to $10$ in increments of .1 in C1. For each of the values in C1 calculate the value of the quadratic polynomial $2x^2 + 4x - 3$ – i.e. substitute the value in each

entry in C1 into this expression – and place these values in C2. Using Minitab commands and the values in C1 and C2 estimate the point in the range from $-10$ to 10 where this polynomial takes its smallest value and what this smallest value is. Using Minitab commands and the values in C1 and C2 estimate the points in the range from $-10$ to 10 where this polynomial is closest to 0.

9. Using the **set** command place values in the range from 0 to 5 using an increment of .01 in C1. Calculate the value of $1 - e^{-x}$ for each value in C1 and place the result in C2. Using Minitab commands find the largest value in C1 where the corresponding entry in C2 is less than or equal to .5. Note that $e^{-x}$ corresponds to the **exponentiate** command (see Appendix B.1) evaluated at $-x$.

10. Using the **set** command place values in the range from $-4$ to 4 using an increment of .01 in C1. Calculate the value of

$$\frac{1}{\sqrt{2\pi}} \, e^{-x^2/2}$$

for each value in C1 and place the result in C2 where $\pi = 3.1415927$. Using **parsums** (see Appendix B.1) calculate the partial sums for C2 and place the result in C3. Multiply C3 times .01. Find the largest value in C1 such that the corresponding entry in C3 is less than or equal to .25.

# Part II

# Minitab for Data Analysis

# Chapter 1

# Looking at Data: Distributions

**New Minitab commands discussed in this chapter**

| boxplot | cdf | describe | histogram | invcdf | mtsplot |
|---------|-----|----------|-----------|--------|---------|
| nscores | pdf | stats | stem and leaf | tally | tsplot |

This chapter of IPS is concerned with the various ways of presenting and summarizing a data set and also introduces the normal distribution. By presenting data we mean convenient and informative methods of conveying the information contained in a data set. There are two basic methods for presenting data, namely graphically and through tabulations. Still it can be hard to summarize exactly what these presentations are saying about the data. So the chapter also introduces various summary statistics that are commonly used to convey meaningful information in a concise way. The normal distribution is of great importance in the theory and application of statistics and it is necessary to gain some facility with carrying out various computations with this distribution.

All of these topics can involve much tedious, error prone calculation, if we were to insist on doing them by hand. An important point is that you should almost never rely on hand calculation in carrying out a data analysis. Not only are there many far more important things for you to be thinking about, as the text discusses, but you are also likely to make an error. On the other hand never blindly trust the computer! Check your results and make sure that they make sense in light of the application. For this a few simple hand calculations can prove valuable. In working through the problems in

47

IPS you should try to use Minitab as much as possible as this will increase your skill with the package and inevitably make your data analyses easier and more effective.

# 1.1   Tabulating and Summarizing Data

If a variable is categorical then we construct a table using the values of the variable and recording the *frequency* (count) of each value in the data and perhaps the *relative frequency* (proportion) of each value in the data as well. These relative frequencies then serve as a convenient summarization of the data.

If the variable is quantitative then we typically *group* the data in some way, i.e. divide the range of the data into nonoverlapping intervals and then record the frequency and proportion of values in each interval. Grouping is accomplished using the **code** command discussed in I.11.1.

If the values of a variable are *ordered* then we can record the *cumulative distribution*, namely the proportion of values less than or equal to each value. Quantitative variables are always ordered but sometimes categorical variables are as well, e.g. when a categorical variable arises from grouping a quantitative variable.

Often it is convenient with quantitative variables to record the *empirical distribution function*, which for data values $x_1, \ldots, x_n$ and at a value $x$ is given by

$$\hat{F}(x) = \frac{\# \text{ of } x_i \leq x}{n}$$

i.e. $\hat{F}(x)$ is the proportion of data values less than or equal to $x$. We can summarize such a presentation via the calculation of a few quantities such as the *first quartile*, the *median* and the *third quartile* or present the *mean* and the *standard deviation*.

We illustrate these presentations using some data from IPS and introduce some new commands.

## 1.1.1   The TALLY Command

The **tally** command tabulates categorical data. Suppose that Newcomb's measurements in Table 1.1 of IPS have been recorded as C1 in a worksheet called **newcomb**. These data range from $-44$ to $40$ (use **min** and **max** in Appendix B.1 to calculate these values). Suppose we decide to group these into the intervals $(-50, 0], (0,20], (20,25], (25,30], (30,35], (35,40]$. Then we want to record the frequencies, relative frequencies, cumulative frequencies and cumulative distribution of this grouped variable. Then

```
MTB >retrieve 'newcomb'
Retrieving worksheet from file:  newcomb.MTW
Worksheet was saved on Sat Feb 21 1998
MTB >code (-50:0) to 1 (1:20) to 2 (21:25) to 3 &
CONT>(26:30) to 4 (31:35) to 5 (36:50) to 6 c1 put in c2
MTB >tally c2;
SUBC>counts;
SUBC>percents;
SUBC>cumcnts;
SUBC>cumpcts.
```

| C2 | Count | Percent | CumCnt | CumPct |
|----|-------|---------|--------|--------|
| 1  | 2     | 3.03    | 2      | 3.03   |
| 2  | 4     | 6.06    | 6      | 9.09   |
| 3  | 17    | 25.76   | 23     | 34.85  |
| 4  | 26    | 39.39   | 49     | 74.24  |
| 5  | 10    | 15.15   | 59     | 89.39  |
| 6  | 7     | 10.61   | 66     | 100.00 |
|    | N= 66 |         |        |        |

accomplishes this using the **code** command discussed in I.11.1 to create an ordered categorical variable C2 taking the values 1, 2, 3, 4, 5 and 6 and then using the **tally** command with the subcommands **counts** to obtain frequencies, **percents** for relative frequencies, **cumcnts** for the cumulative frequency function and **cumpcts** for the cumulative distribution of C2. Any of the subcommands can be dropped.

We can also use **tally** to compute the empirical distribution function of C1 in the **newcomb** worksheet. The commands

```
MTB >sort c1 c3
MTB >tally c3;
SUBC>cumpcnts;
SUBC>store c4 c5.
```

first use the **sort** command, discussed in I.10.1, to sort the data in C1 from smallest to largest and place the results in C3. Then the cumulative distribution is computed for the values in C3 with the unique values in C3 stored in C4 and the cumulative distribution at each of the unique values stored in C5 via the **store** subcommand to **tally**.

The general syntax of the **tally** command is

**tally** $E_1 \ldots E_m$

where $E_1$, ..., $E_m$ are columns of categorical variables and the command is applied to each column. If no subcommands are given, then only frequencies are computed. The **table** command is in essence a more elaborate version of tally and will be discussed in II.3.

In the window environment these results can be obtained using S̲tat ▶ T̲ables ▶ T̲ally and filling in the dialog box appropriately.

## 1.1.2   The DESCRIBE Command

The **describe** command is used with quantitative variables to present a numerical summary of the variable values. These values are in a sense a summarization of the empirical distribution of the variable. For example, in the `newcomb` worksheet,

```
MTB >describe c1
Variable  N  Mean  Median TrMean  StDev SE Mean
C1        66 26.21   27.00  27.40  10.75    1.32
Variable Minimum Maximum    Q1    Q3
C1         -44.00   40.00 24.00 31.00
```

gives the count N, the mean, median, trimmed mean `TrMean` (removes lower 5% and upper 5% of the data and averages the rest), standard deviation, standard error of the mean, minimum, maximum, first quartile `Q1` and third quartile `Q3`.

The **by** subcommand can also be used with **describe** to produce a summary for a variable at each value of another variable. For example, suppose in the `newcomb` worksheet we want a summary of C1 for each of the intervals that we grouped the values of C1 into, i.e. by the values of C2. Then

```
MTB >describe c1;
SUBC>by c2.
```

produces six summaries of C1: one for all the values in (-50,0], one for all the values in (0,20], etc.

The general syntax of the **describe** command is

**describe** $E_1 \ldots E_m$

where $E_1$, ..., $E_m$ are columns of quantitative variables and the command is applied to each column. If a **by** subcommand is also used it is clear that we could generate a considerable amount of output so we need to be careful. Note that a number of the summary statistics can also be computed using the Column Statistics discussed in I.10.3.

In the window environment these summaries can be obtained using Stat ▶ Basic Statistics ▶ Display Descriptive Statistics and filling in the dialog box appropriately.

## 1.1.3    The STATS Command

The **stats** command is like **describe** but a much greater number of summary statistics can be computed and it allows these values to be stored for further use. Of course Column Statistics as discussed in I.10.3 is also available for this purpose, at least for some of the statistics. For example,

```
MTB >stats c1;
SUBC>by c2;
SUBC>glabels c6;
SUBC>mean c7.
MTB >print c6 c7
Data Display
  Row C6        C7
    1  1 -23.0000
    2  2  17.7500
    3  3  23.5294
    4  4  27.8077
    5  5  32.2000
    6  6  37.1429
```

computes the mean of C1 for each value of C2, the subcommand **glabels** stores the unique values of C2 in C6 and the subcommand **mean** stores the respective means in C7. There are various other statistics that can be computed and stored such as **variance**, **skewness** and **kurtosis**. We refer the reader to **help** for a complete listing and description of this command. The **stats** command is not available in earlier versions of Minitab.

In the window environment the command can be accessed via S̲tat ▶ B̲asic Statistics ▶ S̲tore Descriptive Statistics and filling in the dialog box appropriately.

# 1.2    Graphing Data Using Session Commands

One of the most informative ways of presenting data is via a plot. There are many different types of plots within Minitab and which to use depends on the type of variable you have and what you are trying to learn. In this section we describe plotting in the Session window. Such plots are referred to as *character plots*. In the window environment you are also able to generate *higher resolution plots* in a *Graph window*. If the window environment is available to you then you should use these better quality plots but you should still read this section to become familiar with the various plots available.

## 1.2.1    The DOTPLOT Command

The **dotplot** command is used with quantitative variables and produces a plot of each data value as a dot along the $x$-axis so that you get a general idea of the location of the data and how much scatter there is. Actually the data is grouped before plotting and multiple observations in a group are stacked over the $x$-axis with a colon representing two observations. The interval between successive tick ($+$) marks on the $x$-axis is divided into 10 equal length subintervals for the grouping. Typically one also looks for points that are far from the main scatter of points as these may be identified as *outliers* and, as such, deleted from the data set for subsequent analysis. For example, for the `newcomb` worksheet discussed in II.1.1,

    MTB >dotplot c1

produces a character plot that looks something like Figure 1.1. Actually the graph produced on the screen or in the Session window is more rudimentary as Figure 1.1 was produced using the higher resolution plotting capabilities of Minitab, but it is similar.

The general syntax of the dotplot command is

**dotplot** $E_1 \ldots E_m$

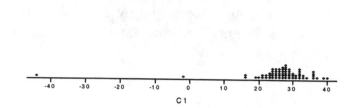

Figure 1.1: Dotplot for C1 in the `newcomb` worksheet.

where $E_1$, ..., $E_m$ are columns and a dotplot is produced for each. There are a number of subcommands available. The **same** subcommand ensures the scales of the dotplots are the same for each column. The **by** subcommand allows plotting of a variable by the values of another variable with all plots having the same scale. The **increment** subcommand allows for control of the distance between the tick marks and **start** and **end** allow you to specify where the dotplot should begin and end. For example,

```
MTB >dotplot c1;
SUBC>increment=5;
SUBC>start=20 end=35.
```

puts the tick marks 5 units apart, starts the plot at 20 and ends it at 35, so some points are not plotted in this case.

## 1.2.2   The HISTOGRAM Commands

The **histogram** command produces *histogram* plots of quantitative variables. A histogram is a plot where the data are grouped into intervals and over each such interval a bar is drawn of height equal to the frequency of data values in that interval or of height equal to the relative frequency (proportion) of data values in that interval or of height equal to the *density* of points in that interval, i.e. the proportion of points in the interval divided by the length of the interval. In the session environment only frequency histograms are available.

As with the **dotplot** command the grouping intervals, and thus the appearance of the histogram, are determined by Minitab unless you use the **increment** and **start** and **end** subcommands. For example, using the newcomb worksheet,

```
MTB >histogram c1;
SUBC>increment=10;
SUBC>start=-50 end=50.
Histogram of C1 N = 66
```

```
Midpoint        Count
 -50.00            0
 -40.00            1 *
 -30.00            0
 -20.00            0
 -10.00            0
   0.00            1 *
  10.00            0
  20.00           16 ****************
  30.00           41 *****************************************
  40.00            7 *******
```

produces a character graphics histogram in the Session window where the data is grouped into intervals of length 10 with midpoints starting at $-50$ and ending at 50. You can also use the **by** subcommand to produce a number of histograms for a variable where the values are first divided into subsets by the value of another variable. You can also produce multiple histograms for a number of columns using a command of the form

**histogram** $E_1 \ldots E_m$

where $E_1$, ..., $E_m$ correspond to columns. The subcommand **same** ensures that all of these histograms have the same scale.

## 1.2.3  The STEM-AND-LEAF Command

Stem-and-leaf plots are similar to histograms and are produced by the **stem-and-leaf** command. These plots are also referred to as *stemplots* as in IPS. For example, using the newcomb worksheet,

```
MTB >stem and leaf c1;
```

```
SUBC>increment=10.
Stem-and-leaf of C1 N = 66
Leaf Unit = 1.0
   1   -4 4
   1   -3
   1   -2
   1   -1
   2   -0 2
   2    0
   5    1 669
 (41)   2 01122333444445555566666777777888888899999
  20    3 00011222223334666679
   1    4 0
```

produces a stem-and-leaf plot of the values in C1 where the subcommand **increment** determines the distance between stems in the plot. The first column gives the *depths* for a given stem, i.e. the number of observations on that line and below it or above it depending on whether or not the observation is below or above the median. The row containing the median is enclosed in parentheses ( ) and the depth is only the observations on that line. If the number of observations is even and the median is the average of values on different rows, then parentheses do not appear. The second column gives the *stems*, as determined by Minitab, and the remaining columns give the ordered *leaves* where each digit represents one observation. The *Leaf Unit* determines where the decimal place goes after each leaf. So in this example the first observation is $-44.0$ while it would be $-4.4$ if the Leaf Unit were .1. Multiple stem-and-leaf plots can be carried out for a number of columns simultaneously and also for a single variable using the **by** subcommand. The subcommand **trim** is also available and this removes outliers from the plot as determined by the *inner fences* discussed in the next section for the **boxplot** command.

## 1.2.4 The BOXPLOT Command

Boxplots are useful summaries of a quantitative variable and are obtained using the **boxplot** command. Boxplots are used to provide a graphical notion of the location of the data and its scatter in a concise and evocative way. For example, in the **newcomb** worksheet,

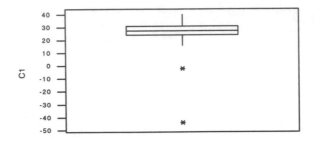

Figure 1.2: Boxplot for C1 in the **newcomb** worksheet.

MTB >boxplot c1

produces the boxplot shown in Figure 1.2. Actually this is a high resolution graphic but the character graph boxplot is similar although laid out horizontally. The line in the center of the box is the median. The line below the median is the first quartile, also called the *lower hinge,* and the line above is third quartile, also called the *upper hinge.* The difference between the third and first quartile is called the *interquartile range* or IQR. The vertical lines from the hinges are called *whiskers* and these run from the hinges to the *adjacent values.* The adjacent values are given by the greatest value less than or equal to the *upper limit* (the third quartile plus 1.5 times the IQR) and by the least value greater than or equal to the *lower limit* (the first quartile minus 1.5 times the IQR). The upper and lower limits are also referred to as the *inner fences.* The *outer fences* are defined by replacing the multiple 1.5 in the definition of the inner fences by 3.0. Values between the inner and outer fences are plotted with a * and values beyond the outer fences are plotted with a O (in high resolution plots all of these values are plotted with a * and are called *outliers*). Multiple boxplots of several columns can be obtained via the command

**boxplot** $E_1 \ldots E_m$

where $E_1$, ..., $E_m$ correspond to columns. Also multiple boxplots of a single variable are available using the **by** subcommand to subset on another variable and all the boxplots are on the same scale.

## 1.2.5   The TSPLOT and MTSPLOT Commands

Often data are collected sequentially in time. In such a context it is instructive to plot the values of quantitative variables against time in a time series plot. For this we use the **tsplot** command. For example, in the `newcomb` worksheet the command

```
MTB >tsplot c1
```

produces a plot similar to the high resolution plot shown in Figure 1.3 in high resolution graphics. This presumes that the first value in C1 is the first value recorded, the second is the second recorded, etc. The time intervals between points are also assumed to be equal and labeled 1, 2, etc. If the measurements are made periodically, then the period placed before the column ensures that the numbers on the $x$-axis are labeled accordingly. For example, suppose that the observations in C1 were made monthly. Then the command

```
MTB >tsplot 12 c1
```

ensures that the tick marks along the $x$-axis are in multiples of 12 and this makes it easy to look at the data in years. The **origin** subcommand controls the label associated with the first measurement. For example, `origin = 20;` starts the labels on the $x$-axis at 20. The **increment** and **start** and **end** subcommands allow you to control the appearance of the plot along the $y$-axis. For example,

```
MTB >tsplot c1;
SUBC>increment=50;
SUBC>start=-50 end =50.
```

puts the first tick mark at $-50$, the second at 0 and the third at 50 on the $y$-axis. The **tstart** subcommand allows you to start the time series at any value. For example, `tstart = 10;` omits the first 9 observations from the plot.

If you have several time series all observed at the same times, then they can all be plotted on the same axes using the **mtsplot** command. The syntax of this command is

**mtsplot** $V_1$ $E_1 \ldots E_m$

where $V_1$ is the period, which can be omitted, and $E_1, ..., E_m$ are the columns containing the time series. The subcommands for **mtsplot** are the same as for **tsplot**.

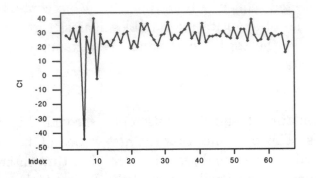

Figure 1.3: Time series plot for C1 in the `newcomb` worksheet.

## 1.3   Graphing Data in a Graph Window

Higher resolution graphics are available in newer versions of Minitab such as Version 12 and they are well worth it if this facility is available to you. All our comments in this section apply only to the window environment. While we discuss the session versions of the commands here plots are also available from Graph on the menu bar, clicking on the relevant plot and then filling in the dialog box appropriately. In many ways using the menu commands rather than the session commands is much easier for graphics. For example dotplots are available via the menu command Graph ▶ Dotplot and stem-and-leaf plots are available via Graph ▶ Stem-and-Leaf.

There are many features of plotting in the window environment that we will not describe. For example, there are many graphical editing capabilities which allow you to add features such as titles or legends. Some of these features are accessed via Graph ▶ Layout. We refer the reader to **help** for more details on these features.

You can toggle back and forth between producing character graphics in the Session window and producing higher resolution graphics in a *Graph window*. The session command **gstd** ensures that the graphics will appear as character graphics in the Session window until the command **gpro** is given and then the graphics will be plotted as higher resolution graphics in Graph windows.

In some older versions of Minitab high resolution graphics were produced by commands such as **ghistogram** and **gplot**. As these are now obsolete we do not discuss these here but refer the reader to **help** if you are using such a version of Minitab.

## 1.3.1 Histograms

The **histogram** command works a little differently with higher resolution graphics. For example, with the `newcomb` worksheet,

```
MTB >histogram c1;
SUBC>midpoints -45 -30 -15 0 15 30 45;
SUBC>density.
```

produces the histogram in Figure 1.4. Here the subcommand **midpoints** specifies that there will be 7 equal length intervals for grouping the data with their respective midpoints at $-45$, $-30$, 15, 0, 15, 30 and 45. The subcommand **density** ensures that the heights of the bars equal the proportion of data values in the interval divided by the length of the interval – 15 in this case – so that the area under the histogram is 1. Rather than **midpoints** we could have used the subcommand **cutpoints** which specifies the ends of the subintervals. The advantage with **cutpoints** is that subintervals of unequal lengths can be specified. If neither of these subcommands is used, then Minitab chooses the endpoints in a nice way for the plot. The number of subintervals can be specified with the subcommand **nintervals**. Rather than using **density** we could have used the subcommands **frequency** or **percent** which respectively ensure that the heights of the bar lines equal the frequency and relative frequency of the data values in the interval. Also the **cumulative** subcommand is available so that the bars represent all the values less than or equal to the endpoint of an interval. These plots can also be accessed via the menu command Graph ▶ Histogram and filling in the dialog box.

## 1.3.2 Boxplots

Boxplots perform as described in II.1.2.4 with the exception of the **by** subcommand which is not available with higher resolution plots. Rather the command

**boxplot** $E_1 * E_2$

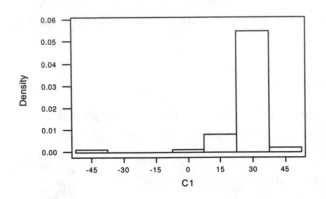

Figure 1.4: A histogram for C1 in the `newcomb` worksheet.

plots boxplots for column $E_1$ by the values in column $E_2$. The menu command Graph ▶ Boxplots is also available for these plots.

## 1.3.3   Time Series Plots

There are many more features associated with time series plots in the window environment than with **tsplot** in the session environment and the subcommands available are somewhat different. We refer the reader to **help**. This command can be accessed via Graph ▶ Time Series Plot and filling in the dialog box accordingly.

## 1.3.4   Bar Charts

In the window environment it is also possible to produce various charts. For example,

> **chart** $E_1$

produces a *bar chart* for the values in column $E_1$; each distinct value of C1 is plotted along the $x$-axis simply as a categorical value, not as a quantitative value, and a bar of height equal to the number of times that value occurs in the variable is drawn. A bar chart is a good way to plot categorical variables. The command

**chart** $E_1 * E_2$

produces a bar chart where the height of the bar equals the sum of the values of the quantitative variable in $E_1$ by the values in the categorical variable in $E_2$. For example, $E_1$ might contain the sales figures for various stores and $E_2$ a categorical variable indicating which area the stores are located in. The above command produces a bar chart of total sales by area. There are various functions other than sum that can be used, such as mean and median. To access this and other features use Graph ▶ Chart and fill in the dialog box appropriately.

### 1.3.5   Pie Charts

A *pie chart* is a disk divided up into wedges where each wedge corresponds to a unique value of a variable and the area of the wedge is proportional to the relative frequency of the value it corresponds to. Pie charts can be obtained via Graph ▶ Pie Chart and there are various features available in the dialog box that can be used to enhance these plots. Pie charts are a common method for plotting categorical variables.

## 1.4   The Normal Distribution

It is important in statistics to be able to do computations with the normal distribution. As noted in IPS the equation of the *density curve* for the normal distribution with mean $\mu$ and standard deviation $\sigma$ is given by

$$\frac{1}{\sqrt{2\pi}} e^{-\frac{1}{2}\left(\frac{z-\mu}{\sigma}\right)^2}$$

where $z$ is a number. We refer to this as the $N(\mu, \sigma)$ density curve. Also of interest is the area under the density curve from $-\infty$ to a number $x$, i.e. the area between the graph of the $N(\mu, \sigma)$ density curve and the interval $(-\infty, x]$. As noted in IPS this is a value between 0 and 1. Sometimes we specify a value $p$ between 0 and 1 and then want to find the point $x_p$ such that $p$ of the area for the $N(\mu, \sigma)$ density curve lies in $(-\infty, x_p]$. The point $x_p$ is called the *pth percentile* of the $N(\mu, \sigma)$ density curve.

Sometimes we are given a mean $\mu$ and a standard deviation $\sigma$ and then asked to standardize a variable $x$ whose values are in some column, i.e., produce the new variable $z = \frac{x-\mu}{\sigma}$. These arithmetical operations can be carried out using the **let** command as described in I.10.1.

## 1.4.1   The PDF Command

*gives y value on curve.*

Suppose that we want to evaluate the $N(\mu, \sigma)$ density curve at a value $x$. For this we use the **pdf** command which stands for *probability density function*. For example, the command

```
MTB >set c1
DATA>-3:3/.01
DATA>end
MTB >pdf c1 c2;
SUBC>normal mu=-.5 sigma=1.2.
```

puts the values between $-3$ and $3$ in increments of .01 in C1 using the **set** command. Then the **pdf** command with the **normal** subcommand calculates the $N(-.5, 1.2)$ density curve at each of these values and puts the outcomes in the corresponding entries in C2. Note that with the **normal** subcommand we must also specify the mean and the standard deviation via **mu** and **sigma**.

The general syntax of the **pdf** command with the **normal** subcommand is

**pdf**  $E_1 \dots E_m$ into $E_{m+1} \dots E_{2m}$;
**normal mu** $= V_1$ **sigma** $= V_2$.

where $E_1$, ..., $E_m$ are columns or constants containing numbers and $E_{m+1}$, ..., $E_{2m}$ are the columns or constants that store the values of the $N(\mu, \sigma)$ density curve at these numbers and $V_1 = \mu$ and $V_2 = \sigma$. If no storage is specified then the values are printed.

In the window environment this command is available via Calc ▶ Probability Distributions ▶ Normal and filling in the dialog box appropriately.

## 1.4.2   The CDF Command

*gives area under ttt*

Suppose that we want to evaluate the area under $N(\mu, \sigma)$ density curve over the interval $(-\infty, x]$. For this we use the **cdf** command which stands for *cumulative distribution function*. For example, the commands

```
MTB >set c1
DATA>-3:3/.01
```

```
DATA>end
MTB >cdf c1 c2;
SUBC>normal mu=-.5 sigma=1.2.
```

put the values between $-3$ and $3$ in increments of .01 in C1 using the **set** command. Then the **cdf** command, with the **normal** subcommand, calculates the area under the $N(-.5, 1.2)$ density curve over $(-\infty, x]$, where $x$ takes on each of the values in C1, and puts the outcomes in the corresponding entries in C2.

The general syntax of the **cdf** command with the **normal** subcommand is

> **cdf** $E_1 \ldots E_m$ into $E_{m+1} \ldots E_{2m}$;
> **normal mu** $= V_1$ **sigma** $= V_2$.

where $E_1$, ..., $E_m$ are columns or constants containing numbers and $E_{m+1}$, ..., $E_{2m}$ are the columns or constants that store the values of the area under $N(\mu, \sigma)$ density curve over the interval from $-\infty$ to these numbers and $V_1 = \mu$ and $V_2 = \sigma$. If no storage is specified then the values are printed.

In the window environment this command is available via Calc ▶ Probability Distributions ▶ Normal and filling in the dialog box appropriately.

### 1.4.3   The INVCDF Command

Suppose that we want to evaluate percentiles for the $N(\mu, \sigma)$ density curve. For this we use the **invcdf** command which stands for *inverse cumulative distribution function*. For example, the commands

```
MTB >set c1
DATA>0:1/.05
DATA>end
MTB >invcdf c1 c2;
SUBC>normal mu=-.5 sigma=1.2.
```

put the values between 0 and 1 in increments of .05 in C1 using the **set** command. Then the **invcdf** command, with the **normal** subcommand, calculates the percentiles for the $N(-.5, 1.2)$ density curve at each of these values and puts the outcomes in the corresponding entries in C2. Note that the first element in C2 is * – i.e. it is missing – because the 0*th* percentile is $-\infty$.

The general syntax of the **invcdf** command with the **normal** subcommand is

**invcdf** $E_1 \ldots E_m$ into $E_{m+1} \ldots E_{2m}$;
**normal mu** = $V_1$ **sigma** = $V_2$.

where $E_1$, ..., $E_m$ are columns or constants containing numbers between 0 and 1 and $E_{m+1}$, ..., $E_{2m}$ are the columns or constants that store the values of the percentiles of the $N(\mu, \sigma)$ density curve at these numbers and where $V_1 = \mu$ and $V_2 = \sigma$. If no storage is specified, then the values are printed.

In the window environment this command is available via Calc ▶ Probability Distributions ▶ Normal and filling in the dialog box appropriately.

## 1.4.4   Normal Quantile Plots

Some statistical procedures require that we assume that values for some variables are a sample from a normal distribution. A *normal quantile plot* is a *diagnostic* that checks for the reasonableness of this assumption. Note that *quantile* means the same as percentile. To create such a plot we use the **nscores** and **plot** commands. For example, for the `newcomb` worksheet, the commands

```
MTB >nscores c1 c2
MTB >plot c2 c1
```

produce a normal quantile plot of the data in C1 something like the high resolution plot that is shown in Figure 1.5. Of course the plot should be like a straight line and it is not in this case. See IPS for further discussion of this example.

The **plot** command will be discussed much more extensively in II.2. The **nscores** (*normal scores*) command relies on some concepts that are beyond the level of this course and so we do not discuss this further.

Actually the high resolution plot shown in Figure 1.5, which is available in the window environment, is produced by the commands

```
MTB >nscores c1 c2
MTB >plot c2*c1
```

and of course we should have previously used the command **gpro** to ensure high resolution graphics. Alternatively we can use the menu command Graph ▶ Probability Plot and then fill in the dialog box appropriately.

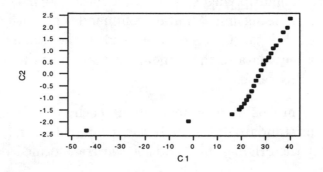

Figure 1.5: Normal quantile plot for C1 in the `newcomb` worksheet.

## 1.5   Exercises

When the data for an exercise come from an exercise in IPS, the IPS exercise number is given in parentheses ( ). All computations in these exercises are to be carried out using Minitab and the exercises are designed to ensure that you have a reasonable understanding of the Minitab material in this chapter. More generally you should be using Minitab to do all the computations and plotting required for the problems in IPS.

1. Using Newcomb's measurements in Table 1.1 of IPS create a new variable by grouping these values into three subintervals $(-50, 0)$, $[0, 20)$, $[20, 50)$. Calculate the frequency distribution, the relative frequency distribution and the cumulative distribution of this ordered categorical variable.

2. Using the data in Example 1.5 of IPS on the amount of money spent by shoppers in a supermarket print the empirical distribution function. From this determine the first quartile, median and third quartile. Also use the empirical distribution function to compute the $10th$ and $90th$ percentiles.

3. (1.23) Use Minitab commands for the stemplot and the histogram. Use Minitab commands to compute a numerical summary of this data and justify your choices.

4. (1.24) Transform the data in this problem by subtracting 5 from each value and then multiplying by 10. Calculate the means and standard deviations, using any Minitab commands, of both the original and transformed data. Compute the ratio of the standard deviation of the transformed data to the standard deviation of the original data. Comment on this value.

5. (1.27) Transform this data by multiplying each value by 3. Compute the ratio of the standard deviation to the mean (called the *coefficient of variation*) for the original data and for the transformed data. Justify the outcome.

6. (1.38) Use Minitab to draw time plots for the mens and women's winning times in the Boston marathon on a common set of axes.

7. For the $N(6, 1.1)$ density curve compute the area between the interval (3,5) and the density curve. What number has 53% of the area to the left of it for this density curve?

8. Use Minitab commands to verify the 68-95-99.7 rule for the $N(2, 3)$ density curve.

9. Calculate and store the values of the $N(0, 1)$ density curve at each value in $[-3, 3]$ using an increment of .01. Put the values in the interval $[-3, 3]$ in C1 and the values of the density curve in C2. Using the command `plot C2 C1` plot the density curve. Comment on the shape of this curve.

10. Use Minitab commands to make the normal quantile plot presented in Figures 1.32 of IPS.

# Chapter 2

# Looking at Data: Relationships

**New Minitab commands discussed in this chapter**

 correlate   lplot   mplot   plot   regress   table
 tplot

In this chapter Minitab commands are described that permit the analysis of relationships among two variables. The methods are different depending on whether or not both variables are quantitative, both variables are categorical or one in is quantitative and the other is categorical. Graphical methods are very useful in looking for relationships among variables and we examine various plots for this.

## 2.1 Relationships Between Two Quantitative Variables

### 2.1.1 The PLOT and Related Commands

A *scatter plot* of two quantitative variables is a very useful technique when looking for a relationship between two variables. By a scatter plot we mean a plot of one variable on the $y$-axis against the other variable on the $x$-axis. For example, consider Example 2.4 in IPS where we are concerned with the relationship between the length of the femur and the length of the humerus for an extinct species. Suppose that we have read in the data so that the

67

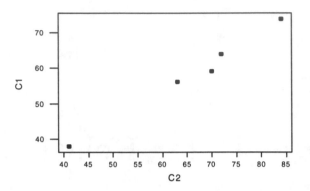

Figure 2.1: Scatter plot of femur length (C1) versus humerus length (C2) of Example 2.4 in IPS.

measurements of femur length are in C1 and the lengths of the humerus are in C2. Then the command

```
MTB > plot c1 c2
```

produces a character plot very like the high resolution plot presented in Figure 2.1. Note that the first column C1 is plotted along the $y$-axis and the second column C2 is plotted along the $x$-axis. There is a generally positive relationship between the two variables just as we would expect.

There are various subcommands that can be used with **plot**. The **title** subcommand allows you to add titles to your plots placed above the plot. For example, the subcommand

```
title = 'Femur length versus humerus length';
```

places the title `Femur length versus humerus length` centered above the plot. You can also label the $x$ and $y$ axes in any way you wish using the **xlabel** and **ylabel** subcommands via `xlabel='text1';` and `ylabel = 'text2';` where `text1` and `text2` are replaced by the labels of your choice. Rather than using * to plot the points another symbol can be chosen using the **symbol** subcommand via `symbol = 's';` where `s` is a symbol of your choosing. The appearance of the $x$- and $y$-axes can be controlled by the **xincrement**, **yincrement**, **xstart**, **ystart** and **end** subcommands. For example,

```
MTB > plot c1 c2;
SUBC> xincrement=15;
SUBC> yincrement=5;
SUBC> xstart= 60 end=80;
SUBC> ystart=50 end=60.
```

puts the tick (+) marks 15 units apart on the $x$-axis, 5 units apart on the $y$-axis, and excludes all points for which the femur length is not between 60 and 80 and the humerus length is not between 50 and 60. So in fact there are only two points plotted. Minitab will override the tick mark settings if the plot becomes too large to present the data.

The **mplot** (multiple plot) command is used when you want several scatter plots to appear on the same set of axes. The syntax for this command is

**mplot** $E_1$ vs $E_2$ ... $E_{m-1}$ vs $E_m$

which results in column $E_1$ being plotted against column $E_2$, column $E_3$ against column $E_4$, etc. The subcommands for **mplot** are the same as those for **plot**.

The **lplot** (labeled plot) command produces scatter plots where the points are labeled. For example, suppose that C3 contains the values 1, 1, 1, 2, 2. Then

```
MTB > lplot c1 c2 c3
```

produces a scatter plot where the first three observations are plotted using the symbol A and the last two observations with the symbol B.

If you have three variables it is possible to obtain a (pseudo) three-dimensional scatter plot using the **tplot** (three-dimensional plot) command. The general syntax of this command is

**tplot** $E_1$ $E_2$ $E_3$

where $E_1$, $E_2$ and $E_3$ are columns of quantitative data corresponding to the $y$-, $x$- and $z$-axes respectively. The **contour** command is also available for *contour plots* of three-dimensional data. A contour plot of the variables $x, y$ and $z$ is a set of scatter plots of $x$ and $y$ at a number of values of $z$ all using the same $x$- and $y$- axes. Each scatter plot of $x$ and $y$ corresponds to a constant height for $z$ and hence the term contour. We refer the reader to **help** for more information on this.

## 2.1.2   Higher Resolution Scatter Plots

To produce the scatter plot in Figure 2.1 in the window environment the relevant higher resolution plot command is

```
MTB > plot c1*c2
```

provided that we have previously issued the **gpro** command in the Session window. Alternatively we can use the menu command Graph ▶ Plot and then fill in the dialog box appropriately. Plots labeling the individual points are available using Graph ▶ Plot ▶ Annotation ▶ Data Label.

The technique of *brushing* is available where after obtaining the plot to see which observations (rows) the points correspond to. This is helpful in identifying which points correspond to outliers. Brushing is accessed from the toolbar just below the menu bar by clicking on the brush when the Graph window is active.

Rather than just plotting the points in a scatter plot you can add *connection lines* (join the points with lines), add *projection lines* (drop a line from each point to the $x$-axis), add *areas* (fill in the area under a polygon joining the points). Also you can employ the scatter plot smoother *lowess* to plot a piecewise linear continuous curve through the scatter of points. These features are available via Graph ▶ Plot ▶ Display ▼ where ▼ denotes the arrow beside Display that you are to click on.

There are a number of additional plots available in the window environment. For example, a *marginal plot* of two variables is a scatter plot of one variable against the other where in addition histograms, dotplots or boxplots are plotted along the sides of the scatter plot for each variable. These are available via the menu command Graph ▶ Marginal Plot. *Draftsman plots* allow you to produce a number of scatter plots in a rectangular array so that they can be compared. For example, you may want to plot C1 against C3, C2 against C3, C1 against C4, C2 against C4 and see all of these in a common plot. This capability is available via the menu command Graph ▶ Draftsman Plot and then filling in the dialog box. *Matrix plots* provide a mechanism for placing a number of scatter plots in a rectangular array or matrix so that they can be directly compared or examined for relationships. For example, in a worksheet where C1, C2 and C3 correspond to quantitative variables, the command

```
MTB > matrixplot c1 c2 c3
```

produces six scatter plots by plotting each of the columns against the others.

Matrix plots are also available via the menu command Graph ▶ Matrix Plot. Also three-dimensional scatter plots are available via Graph ▶ 3D Plot and contour plots via Graph ▶ Contour Plot.

Any higher resolution plot can be saved into a file with the file extension .mgf using File ▶ Save Graph As and supplying a file name. A saved graph can be reopened using File ▶ Open Graph.

### 2.1.3   The CORRELATE Command

While a scatter plot is a convenient graphical method for assessing whether or not there is any relationship between two variables, we would also like to assess this numerically. The *correlation coefficient* provides a numerical summarization of the degree to which a linear relationship exists between two quantitative variables and this can be calculated using the **correlate** command. For example, for the data of Example 2.4 in IPS and depicted in Figure 2.1, the command

```
MTB > correlate c1 c2
Correlation of C1 and C2 = 0.994, P-Value = 0.001
```

calculates the correlation coefficient between these two variables as .994 which is quite high. For now we ignore the number recorded as `P-Value`. The general syntax of the **correlate** command is given by

**correlate** $E_1 \ldots E_m$

where $E_1$, ..., $E_m$ are columns corresponding to numerical variables and a correlation coefficient is computed between each pair. This gives $m(m-1)/2$ correlation coefficients. The subcommand **nopvalues** is available if you want to suppress the printing of $P$-values.

In the window environment correlation coefficients can also be computed using Stat ▶ Basic Statistics ▶ Correlation.

### 2.1.4   The REGRESS Command

Regression is another technique for assessing the strength of a linear relationship existing between two variables and it is closely related to correlation. For this we use the **regress** command. Actually regression analysis applies to the analysis of many more variables than just two. We will fully discuss the **regress** command in II.10 and II.11.

As noted in IPS the regression analysis of two quantitative variables involves computing the least-squares line $y = a + bx$ where one variable is taken to be the response variable $y$ and the other is taken to be the explanatory variable $x$. Note that the least squares line is different depending upon which choice is made. For example, for the data of Example 2.4 in IPS and plotted in Figure 2.1 letting femur length be the response and humerus length be the explanatory variable, the command

```
MTB > regress c1 1 on c2
The regression equation is
C1 = 3.70 + 0.826 C2
Predictor      Coef     StDev       T      P
Constant       3.701    3.497    1.06  0.368
C2             0.82574  0.05180  15.94  0.001
S = 1.646 R-Sq = 98.8% R-Sq(adj) = 98.4%
Analysis of Variance
Source            DF      SS       MS        F        P
Regression         1   688.67   688.67   254.10   0.001
Residual Error     3     8.13     2.71
Total              4   696.80
```

gives the least-squares line as $y = 3.70 + 0.826x$; i.e. $a = 3.70$ and $b = 0.826$, which we also see under the **Coef** column in the first table. Also we obtain the value of the square of the correlation coefficient as **R-Sq = 98.8%**. We will discuss the remaining output from the **regress** command in II.10.

Using the subcommands **pfits**, **residual** and **sresidual** we can calculate and store *fitted values*, *residuals* and *standardized residuals* respectively. For example,

```
MTB > regress c1 1 on c2;
SUBC> fits c3;
SUBC> residuals c4;
SUBC> sresiduals c5.
```

carries out the regression analysis above but also stores the fitted values in C3 – the values of $\hat{y} = 3.70 + 0.826x$ at each $x$ value in C2 – stores the residuals $y - \hat{y}$ in C4 and stores the standardized residuals – each residual divided by its standard error – in C5.

Typically we would like to plot the fitted line together with the original points. Then the commands

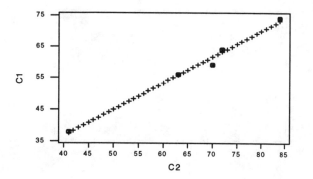

Figure 2.2: Least-squares line together with points in Example 2.4 of IPS.

```
MTB > set c6
DATA> 41:84
DATA> end
MTB > let c7=3.70 + 0.826*C6
MTB > mplot c1 vs c2 c7 vs c6
```

put the points 41, 42, ..., 84 into C6, then evaluate the least-squares line at each of these points and store this in C7 and finally plot the scatter plot of C1 against C2 on the same graph as the scatter plot of C7 against C6 which of course approximates the line. Different symbols are used for each scatter plot. In Figure 2.2 we give a high resolution graph of this plot using the symbol ⊠ for the points and the symbol + for the line. Observe the closeness of the fit. Often we also want to plot the residuals as well to check, for example, that the assumption of a linear relationship approximately holding between the variables makes sense. Then

```
MTB > plot c5 c2
```

plots the standardized residuals against the explanatory variable humerus — $x$ length in a scatter plot as depicted in the high resolution graphic presented in Figure 2.3. There are only a few points in this example but the residual plot looks reasonable.

Various *regression diagnostics* are also available as subcommands. For example,

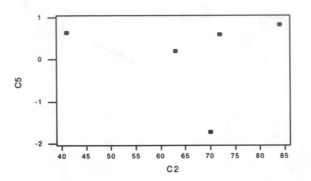

Figure 2.3: Scatter plot of standardized residuals for Example 2.4 in IPS.

```
MTB > regress c1 1 on c2;
SUBC> dfits c6;
SUBC> tresiduals c7.
```

puts the *dffits* in C6 using the **dfits** (note the different spelling) subcommand and puts the *studentized residuals* in C7 using the **tresiduals** subcommand. For a discussion of these concepts see Chapter 2 of IPS. As recommended in IPS a normal quantile plot of the studentized residuals is obtained via the commands

```
MTB > nscores c7 c8
MTB > plot c8 c7
```

and a look at this shows a marked deviation from a straight line.

Sometimes transformations of the variables are appropriate before we carry out a regression analysis. This is accomplished in Minitab using the arithmetical and mathematical operations discussed in I.10.1 and I.10.2. In particular when a residual plot looks bad sometimes this can be fixed by transforming one or more of the variables using a simple transformation such as **loge** (see Appendix B.1).

In the window environment regression analysis can be carried out using Stat ▶ Regression ▶ Regression and filling in the dialog box appropriately. Also the regression as well as a plot with the least-squares line overlaid can be obtained via Stat ▶ Regression ▶ Fitted Line Plot and residual plots obtained using $\underline{S}$tat ▶ $\underline{R}$egression ▶ $\underline{R}$esidual Plots provided you have saved the residuals.

## 2.2 Relationships Between Two Categorical Variables

The relationship between two categorical variables is typically assessed by cross-tabulating the variables in a table. For this the **table** command is available. We illustrate using an example where each categorical variable takes two values but of course each variable can take a number of values and they need not be the same for each categorical variable.

Suppose that we have collected data on courses being taken by students and have recorded a 1 in C2 if the student is taking Statistics and a 0 if not and if the student is taking Calculus a 1 is recorded in C3 and a 0 otherwise. Also we have recorded the student number in C1. These data for 10 students follow.

```
MTB > print c1-c3
  Row        C1        C2        C3
    1     12389         1         0
    2     97658         1         0
    3     53546         0         1
    4     55542         0         1
    5     11223         1         1
    6     77788         0         0
    7     44567         1         1
    8     32156         1         0
    9     33456         0         1
   10     67945         0         1
```

### 2.2.1 The TABLE Command

We cross-tabulate the data in C2 and C3 using the **table** command:

```
MTB > table c2 c3

 Rows:  C2     Columns:  C3
```

```
                    0           1          All

    0               1           4           5
    1               3           2           5
   All              4           6          10

       Cell Contents --
                       Count
```

produces a table that reveals there is 1 student taking neither Statistics nor Calculus, 4 students taking Calculus but not Statistics, 3 students taking Statistics but not Calculus and 2 students taking both subjects. The row marginal totals are produced on the right and the column marginal totals are produced beneath the table. By default the cell entries in the table are frequencies (counts). Various subcommands are available for other choices. For example, the **totpercents** subcommand in

```
MTB > table c2 c3;
SUBC> totpercents.
```

```
    Rows:  C2        Columns:  C3

                    0           1          All

    0              10.00       40.00       50.00
    1              30.00       20.00       50.00
   All             40.00       60.00      100.00

       Cell Contents --
                     % of Tbl
```

produces a table where each entry is the percentage that cell represents of the total number of observations used to form the table.

To examine the relationship between the two variables we compare the conditional distributions given row using the **rowpercents** subcommand or the conditional distributions given column using the **colpercents** subcommand. For example,

```
MTB > table c2 c3;
```

```
SUBC> rowpercents.

    Rows:   C2          Columns:   C3

                0         1        All

    0        20.00     80.00     100.00
    1        60.00     40.00     100.00
    All      40.00     60.00     100.00


    Cell Contents --
                  % of Row
```

gives the row distributions as 20%, 80% for the first row and 60%, 40% for the second row. So it looks as if there is a strong relationship between the variable indicating whether or not a student takes Statistics and the variable indicating whether or not a student takes Calculus. For example, a student who does not take Statistics is more likely to take Calculus than a student who does not take Calculus. Of course this is not a real data set and it is small at that, so in reality we could expect a somewhat different conclusion.

If you do not want the marginal statistics to be printed, then use the **noall** subcommand. Any cases with missing values are not included in the cross-tabulation. If you want them to be included use the **missing** subcommand and a row or column is printed, whichever is relevant, for missing values. For example, the subcommand

```
SUBC> missing c2 c3;
```

ensures that any cases with missing values in C2 or C3 are also tabulated.

In the window environment tables can be created using Stat ▶ Tables ▶ Cross Tabulation and filling in the dialog box appropriately. Some graphical techniques are also available in this environment. In Figure 2.4 we have plotted the conditional distributions given row in a bar chart using the command Graph ▶ Charts and filling in the dialog box so that Function is Count, the $Y$-variable is C3, the $X$-variable is C2, Display is Bar and under Options we selected Cluster with variable C3 and Total $Y$ to 100% within each $X$-category. The bars for C3 are ordered according to the increasing value of C2. If you would rather there be a single bar for each category of the $X$-variable and this bar be subdivided according to the conditional distribution of the

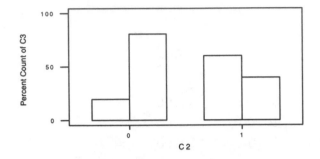

Figure 2.4:  Conditional distributions of columns given row for Example II.2.3.1.

$Y$-variable then, rather than Cluster with variable $Y$, use Stack with variable $Y$. These plots are an evocative way to display the relationship between the variables.

## 2.3   Relationship Between a Categorical Variable and a Quantitative Variable

Suppose now that one variable is categorical and one is quantitative. We treat the situation where the categorical variable is explanatory and the quantitative variable is the response.  The reverse situation is covered in II.15.

To illustrate we use the data in Exercise 2.16 of IPS. Here we have four different colors of insect trap – lemon yellow, white, green and blue – and the number of insects trapped on six different instances of each trap. We have read these data into a worksheet so that C1 contains the trap color with 1 indicating lemon yellow, 2 indicating white, 3 indicating green and 4 indicating blue and in C2 have put the numbers of insects trapped. We then calculate the mean number of insects trapped for each trap using the **table** command

```
MTB > table c1;
SUBC> means c2.
```

```
Rows:   C1

C2
Mean

1   47.167
2   15.667
3   31.500
4   14.833
All 27.292
```

with the **means** subcommand. Also available as subcommands to compute statistics for each cell of the table created by the categorical variable are **medians**, **sums**, **minimums**, **maximums**, **n** (count of the nonmissing values), **nmiss** (count of the missing values), **stdev**, **stats** (equivalent to **n**, **means** and **stdev**) and **data** (lists the data for each cell). In addition there is a subcommand **proportion** with the syntax

**proportion** = V E$_1$;

and this gives the proportion of cases that have the value V in column E$_1$.

It is also a good idea to look at a scatter plot of the quantitative variable versus the categorical variable and to plot the means on this plot as well. Suppose for the data above we have stored the values 1, 2, 3, 4 in C3 and the corresponding means in C4 (recall that the **stats** command can do this directly for you). Then the command

```
MTB > mplot c2 vs c1 c4 vs c3
```

produces a scatter plot of the data together with the means. In Figure 2.5 we have given a high resolution version of this plot with the points denoted by ○ and the means by ●. The character plot, in this case, is in fact not very useful as it has difficulty distinguishing points when they are close together.

Another useful plot is to create side-by-side boxplots. The command
```
MTB > boxplot c2;
SUBC> by c1.
```
creates a boxplot for the data in C2 for each value in C1 and plots them on the same graph.

In the window environment tables can be created using Stat ▶ Tables ▶ Cross Tabulation and filling in the dialog box appropriately. Note that this

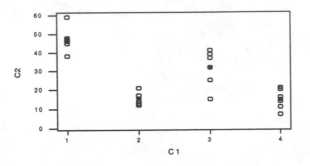

Figure 2.5: Scatter plot of the data in Exercise 2.16 of IPS together with means.

involves using Summaries in the dialog box to choose the cell statistics. Bar charts may also be useful in this context and these are obtained as described in II.1.3.4.

## 2.4 Exercises

When the data for an exercise come from an exercise in IPS, the IPS exercise number is given in parentheses ( ). All computations in these exercises are to be carried out using Minitab and the exercises are designed to ensure that you have a reasonable understanding of the Minitab material in this chapter. More generally you should be using Minitab to do all the computations and plotting required for the problems in IPS.

1. (2.8) Calculate the least-squares line and make a scatter plot of Fuel used against Speed together with the least-squares line. Plot the standardized residuals against Speed. What is the squared correlation coefficient between these variables?

2. (2.10) Make a scatter plot of Rate against Mass and where the points for males and females are labeled differently (use Minitab for the labeling too) and with the least-squares line on it. Hint: Make use of the **stack** command discussed in I.11.11.

3. (2.17) Make a scatter plot of Weight against Pecking Order that includes the means and labels the points according to which pen they correspond to. You may have to manually indicate on the graph where the means are if points overlap.

4. Place the values 1 through 100 with an increment of .1 in C1 and the square of these values in C2. Calculate the correlation coefficient between C1 and C2. Multiply each value in C1 by 10, add 5 and place the results in C3. Calculate the correlation coefficient between C2 and C3. Why are these correlation coefficients the same?

5. Place the values 1 through 100 with an increment of .1 in C1 and the square of these values in C2. Calculate the least-squares line with C2 as response and C1 as explanatory variable. Plot the standardized residuals. If you see such a pattern of residuals what transformation might you use to remedy the problem?

6. (2.40) For the data in this problem numerically verify the algebraic relationship that exists between the correlation coefficient and the slope of the least-squares line.

7. For Example 2.17 in IPS calculate the least-squares line and reproduce Figure 2.21. Calculate the sum of the residuals and the sum of the squared residuals and divide this by the number of data points minus 2. Is there anything you can say about what these quantities are equal to in general?

8. Suppose that the observations in the following table are made on two categorical variables where variable 1 takes 2 values and variable 2 takes 3 values. Using the **table** command cross-tabulate this data in a table of frequencies and in a table of relative frequencies. Calculate the conditional distributions of variable 1 given variable 2. If your version of Minitab allows for bar charts plot the conditional distributions. Is there any indication of a relationship existing between the variables? How many conditional distributions of variable 2 given variable 1 are

there?

| Obs | Var 1 | Var 2 |
|-----|-------|-------|
| 1 | 0 | 2 |
| 2 | 0 | 1 |
| 3 | 0 | 0 |
| 4 | 1 | 0 |
| 5 | 1 | 2 |
| 6 | 0 | 1 |
| 7 | 1 | 2 |
| 8 | 0 | 0 |
| 9 | 0 | 1 |
| 10 | 1 | 1 |

9. Place the values 1 through 10 with an increment of .1 in C1 and place $\exp(-1+2x)$ of these values in C2. Calculate the least-squares line using C2 as the response variable and plot the standardized residuals against C1. What transformation would you use to remedy this residual plot? What is the least-squares line when you carry out this transformation?

10. (2.90) For the table given in this problem use Minitab commands to calculate the marginal distributions and the conditional distributions given field of study. Note you cannot use **table** for this. If the version of Minitab you are using has bar charts available then plot the conditional distributions.

# Chapter 3

# Producing Data

**New Minitab commands discussed in this chapter**
base   random   sample

This chapter is concerned with the collection of data, perhaps the most important step in a statistical problem, as this determincs the quality of whatever conclusions are subsequently drawn. A poor analysis can be fixed if the data collected are good by simply redoing the analysis. But if the data have not been appropriately collected, then no amount of analysis can rescue the study. We discuss Minitab commands that enable you to generate samples from populations and also to randomly allocate treatments to experimental units.

Once data have been collected they are analyzed using a variety of statistical techniques. Virtually all of these involve computing *statistics* that measure some aspect of the data concerning questions we wish to answer. The answers determined by these statistics are subject to the uncertainty caused by the fact that we typically do not have the full population but only a sample from the population. As such we have to be concerned with the variability in the answers when different samples are obtained. This leads to a concern with the *sampling distribution* of a statistic. To assess the sampling distribution of a statistic we make use of a powerful computational tool known as *simulation* which we discuss in this and the following chapter.

Minitab uses computer algorithms to mimic randomness. Still the results are not truly random and in fact any simulation in Minitab can be repeated, with exactly the same results being obtained, using the **base** command. So if you think you might want to repeat your simulation your first step in a simulation is to issue a command of the form

    **base** V

where V is an integer. Then when you want to repeat the simulation you give this command, with the same integer, and provided you use the same simulation commands, you will get the same results. In the window environment this can also be accomplished using Calc ▶ Set Base.

## 3.1   The SAMPLE Command

Suppose that we have a large population of size $N$ and we want to select a sample of $n < N$ from the population. Further we suppose that the elements of the population are ordered; i.e. we have been able to assign a unique number $1, \ldots, N$ to each element of the population. To avoid selection biases we want this to be a random sample; i.e. every subset of size $n$ from the population has the same "chance" of being selected. As discussed in IPS this implies that we generate a *random sample*. We can do this physically using some simple random system such as chips in a bowl or coin tossing, we could use a table of random numbers or, more conveniently, we can use computer algorithms that mimic the behavior of random systems.

For example, suppose there are 1000 elements in a population and we want to generate a sample of 50 from this population without replacement. Then we can use the **sample** command to do this as in

```
MTB > set c1
DATA> 1:1000
DATA> end
MTB > sample 50 C1 C2
MTB > print c2
C2
 441 956   87 736 185 515 883 957 690
 438 205 760 246   16 321 371 493 393
 538 348   70   54 362 492 182 841 287
 277 112 610 890 503 332 413 886 798
 764 584 566 495 547 488 206 557 263
```

```
414 613 618 685 864
```

where we have used the **set** command to put the labels $1, \ldots, 1000$ in C1 and then used the **sample** command to generate the sample and store it in C2. So now we go to the population and select the elements labeled 441, 956, 87, etc. The algorithm that underlies the **sample** command is such that we can be confident that this sample of 50 is like a random sample.

The general syntax of the **sample** command is

**sample** V $E_1 \ldots E_m$ put into $E_{m+1} \ldots E_{2m}$

where V is the sample size $n$ and V rows are sampled from the columns $E_1$, ..., $E_m$ and stored in columns $E_{m+1}$, ..., $E_{2m}$. If we wanted to sample with replacement – i.e. after a unit is sampled it is placed back in the population so that it can possibly be sampled again – then we use the **replace** subcommand. Of course for simple random sampling we do not use the **replace** subcommand. Note that the columns can be numeric or text.

Sometimes we want to generate *random permutations*; i.e., $n = N$ and we are simply reordering the elements of the population. For example, in experimental design suppose we have $N = n_1 + \cdots + n_k$ experimental units and $k$ treatments and we want to allocate $n_i$ applications of treatment $i$ and further we want all possible such applications to be equally likely. Then we generate a random permutation $(l_1, \ldots, l_N)$ of $(1, \ldots, N)$ and allocate treatment 1 to those experimental units labeled $l_1, \ldots, l_{n_1}$, allocate treatment 2 to those experimental units labeled $l_{n_1+1}, \ldots, l_{n_1+n_2}$, etc. For example, if we have 30 experimental units and 3 treatments and we want to allocate 10 experimental units to each treatment then the commands

```
MTB > set c1
DATA> 1:30
DATA> end
MTB > sample 30 c1 c2
MTB > print c2
C2
 13   7 26   8 22 23 28 17   3 25
  9   2 14 29 15 18   6 11 16   5
 12 27   4 30 20 24   1 19 21 10
```

accomplish this. For the treatment allocation you can read the numbers row-wise or column-wise as long as you are consistent. Row-wise is probably best as this is how the numbers are stored in C2 and so you can always refer back to C2 (presuming you save your worksheet) if you get mixed up.

The above examples show how to directly generate a sample from a population of modest size but what happens if the population is huge or it is not convenient to label each unit with a number? For example, suppose we have a population of size 100,000 for which we have an ordered list and we want a sample of size 100. Then more sophisticated techniques need to be used but simple random sampling can still typically be accomplished; see Exercise 3 for a simple method that works in some contexts.

In the window environment random sampling can also be carried out using the menu command Calc ▶ Random Data ▶ Sample from Columns and then filling in the dialog box appropriately.

## 3.2   The RANDOM Command

Once we have generated a sample from a population we measure various attributes of the sampled elements. For example, if we were sampling from a population of humans we might measure each sampled unit's height. The height for the sample unit is now a random quantity that follows the height distribution in the population we are sampling from. For example, if 80% of the people in the population are between 4.5 feet and 6 feet then under *repeated sampling* of an element from the population (with replacement) in the long run 80% of the sampled units will have their heights in this range.

Sometimes we want to sample directly from this population distribution, i.e. generate a number in such a way that under repeated sampling the proportion of values falling in any range agrees with that prescribed by the population distribution. Of course we typically don't know the population distribution as this is what we want to find out about in a statistical investigation. Still there are many instances where we want to pretend that we do know it and simulate from this distribution; e.g. perhaps we want to consider the effect of various choices of population distribution on the sampling distribution of some statistic of interest.

There are computer algorithms that allow us to do this for a variety of different distributions. In Minitab this is accomplished using the **random** command and various subcommands. For example, suppose that we want to simulate the tossing of a fair coin (a coin where head and tail are equally likely as outcomes). Then the command

```
MTB > random 100 c1;
SUBC> bernoulli .5.
```

simulates the tossing of a fair coin 100 times and places the results in C1 where heads is denoted by 1 and tails by 0. The fact that we are simulating coin tossing with heads coded as 1 and tails by 0 is indicated by the **bernoulli** subcommand. The general syntax of the **bernoulli** subcommand is

**bernoulli** V.

where V denotes a value between 0 and 1 and reflects the chance of a 1 occurring; i.e. in repeated sampling from this distribution in the long run a proportion of V will be 1's. For example, bernoulli .7 would simulate coin-tossing from a coin where the chance of a head is .7. In general this distribution is called the *Bernoulli(p)* distribution where $p$ is the probability of getting a head in a single toss; e.g. $p = .7$.

Often a normal distribution with some particular mean and standard deviation is considered a reasonable assumption for the distribution of a measurement in a population. For example,

```
MTB > random 200 c1;
SUBC> normal mu=2.1 sigma=3.3.
```

generates a sample of 200 from the $N(2.1, 3.3)$ distribution.

The general syntax of the **random** command is

**random** V into $E_1 \ldots E_m$

and this puts a sample of size V into each of the columns $E_1$, ..., $E_m$ according to the distribution specified by the subcommand. If no subcommand is provided, then this distribution is taken to be the $N(0, 1)$ distribution.

There are a number of other subcommands specifying distributions, For example,

**integer** $V_1 V_2$

specifies the *uniform distribution* on the integers from $V_1$ to $V_2$; i.e. each integer in this range is equally likely to occur. Also there are many occasions where we want to specify the values that can occur and their chance of occurrence. For example,

**discrete** $E_1 E_2$

uses the values in column $E_1$ as the values to be generated and $E_2$ is a column of the same length that contains values between 0 and 1 that reflect the chance of the corresponding entry in $E_1$ being selected. The values in $E_2$ must be nonnegative and sum to 1 and are called a *probability distribution* on the values in $E_1$. Probability distributions are discussed in Chapter 4 of IPS. For example,

```
MTB > read c1 c2
DATA> 1 .23
DATA> 2 .47
DATA> 3 .30
DATA> end
 3 rows read.
MTB > random 30 c3;
SUBC> discrete c1 c2.
```

places the values to be generated 1, 2 and 3 in C1, with respective probabilities .23, .47 and .30 and then a sample of size 30 is generated from this distribution and stored in C2. For a full list of all the distributions that can be sampled from in Minitab see **help**.

In the window environment random sampling can also be carried out using the menu command Calc ▶ Random Data, clicking on the appropriate distribution and then filling in the dialog box appropriately.

## 3.3   Simulating Sampling Distributions

Once a sample is obtained we compute various statistics based on this data. For example, suppose we flip a possibly biased coin $n$ times and then want to estimate the unknown probability $p$ of getting a head. The natural estimate is $\hat{p}$ the proportion of heads in the sample. We would like to assess the sampling behavior of this statistic in a simulation. To do this we choose a value for $p$, then generate $N$ samples from the Bernoulli distribution of size $n$, for each of these compute $\hat{p}$, then look at the empirical distribution of these $N$ values, perhaps plotting a histogram as well. The larger $N$ is the closer the empirical distribution and histogram will be to the true sampling distribution of $\hat{p}$.

Note that there are two sample sizes here: the sample size $n$ of the original sample the statistic is based on, which is fixed, and the *simulation* sample size $N$, which we can control. This is characteristic of all simulations. Sometimes, using more advanced  analytical techniques, we can determine $N$ so that the sampling distribution of the statistic is estimated with some prescribed accuracy. Some techniques for doing this are discussed in later chapters of IPS. Another method is to repeat the simulation a number of times slowly increasing $N$ until we see the results stabilize. This is sometimes the only way available but caution should be shown as it is easy for simulation results to be very misleading if the final $N$ is too small.

We illustrate a simulation to determine the sampling distribution of $\hat{p}$ when sampling from a *Bernoulli*(.75) distribution. The commands

```
MTB > random 1000 c1-c10;
SUBC> bernoulli .75.
MTB > rmean c1-c10 c11      - prop. 1's
MTB > tally c11;
SUBC> cumpcts.
Summary Statistics for Discrete Variables
  C11 CumPct
  0.3    0.40
  0.4    2.20
  0.5    7.60
  0.6   23.10
  0.7   47.70
  0.8   78.00
  0.9   94.70
  1.0  100.00
```

generate $N = 1000$ samples of size $n = 10$ from the *Bernoulli*(.75) distribution; i.e. we simulated the tossing of this coin 10,000 times, and we placed the results in the rows of columns C1-C10. The proportion of heads $\hat{p}$ in each sample is computed and placed in C11 using the **rmean** command. Note that a mean of values equal to 0 or 1 is just the proportion of 1's in the sample. Finally we used the **tally** command to compute the empirical distribution function of these 1000 values of $\hat{p}$. For example, this says 78% of these values were .8 or smaller and there were no instances smaller than .3. In Figure 3.1 we have plotted a histogram of the 1000 values of $\hat{p}$. Based on $N = 800$ the following empirical distribution was obtained:

```
C11   CumPct
  0.4    1.20
  0.5    7.20
```

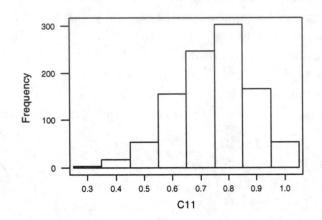

Figure 3.1: Histogram of simulation of $N = 1000$ values of $\hat{p}$ based on a sample of size $n = 10$ from the Bernoulii(.75) distribution.

```
0.6   22.20
0.7   47.80
0.8   78.20
0.9   95.00
1.0  100.00
```

Since these values are reasonably close to those obtained with $N = 1000$ we stopped at $N = 1000$.

In Chapter 5 of IPS we see that the sampling distribution of $\hat{p}$ can be determined exactly; i.e. there are formulas to determine this and we can simulate directly from the sampling distribution, so this simulation can be made much more efficient. In effect this entails using the **binomial** subcommand as in

```
MTB > random 1000 c1;
SUBC> binomial 10 .75.
MTB > let c1=c1/10
```

which also generates $N = 1000$ values of $\hat{p}$ but uses a much smaller number of cells. Still there are many statistics for which this kind of efficiency reduction is not available and, to get some idea of what their sampling distribution is like, we must resort to the more brute force form of simulation of generating directly from the population distribution.

Sometimes more sophisticated simulation techniques are needed to get an accurate assessment of a sampling distribution. Within Minitab there are the macro and exec programming techniques, discussed in Appendix C, that can be applied in such cases. For example, it is clear that if our simulation required the generation of $10^6$ cells, and this is not at all uncommon for some harder problems, then the simulation approach we have described would not work within Minitab as the worksheet would be too large.

## 3.4 Exercises

When the data for an exercise come from an exercise in IPS, the IPS exercise number is given in parentheses ( ). All computations in these exercises are to be carried out using Minitab and the exercises are designed to ensure that you have a reasonable understanding of the Minitab material in this chapter. More generally you should be using Minitab to do all the computations and plotting required for the problems in IPS.

*If your version of Minitab places restrictions such that the value of the simulation sample size N requested in these problems is not feasible, then substitute a more appropriate value. Be aware, however, that the accuracy of your results is dependent on how large N is.*

1. (3.14) Generate a random permutation of the names.

2. (3.27) Use the **sort** command described in I.11.10 to order the subjects by weight. Use the values 1-5 to indicate five blocks of equal length in a separate column and then use the **unstack** command described in I.11.12 to put the blocks in separate columns. Generate a random permutation of each block.

3. Use the following methodology to generate a sample of 20 from a population of 100,000. First put the values 0, 1 in C1 and then the values 0-9 in each of C2-C6. Then use sampling with replacement to generate 50 values from C1 and put the results in C7. Do the same for each of C2-C6 and put the results in C8-C12 (don't generate from these columns simultaneously). Now use the **convert** and **concatenate** commands described in I.11.3 and I.11.2 to create a single column of numbers

using the digits in C7-C12. Pick out the first unique 20 entries as labels for the sample. If you do not obtain 20 unique values, then repeat the process until you do. Why does this work?

4. Suppose you wanted to carry out stratified sampling where there are 3 strata, with the first stratum containing 500 elements, the second stratum containing 400 elements and the third stratum containing 100 elements. Generate a stratified sample with 50 elements from the first stratum, 40 elements from the second stratum and 10 elements from the third stratum (when the strata sample sizes are the same proportion of the total sample size as the strata population sizes are of the total population size this is called *proportional sampling*).

5. Carry out a simulation study with $N = 1000$ of the sampling distribution of $\hat{p}$ for $n = 5, 10, 20$ and for $p = .5, .75, .95$. In particular calculate the empirical distribution functions and plot the histograms. Comment on your findings.

6. Carry out a simulation study with $N = 2000$ of the sampling distribution of the sample standard deviation when sampling from the $N(0, 1)$ distribution based on a sample of size $n = 5$. In particular plot the histogram using cutpoints 0, 1.5, 2.0 2.5, 3.0 5.0. Repeat this for the sample coefficient of variation (sample standard deviation divided by the sample mean) using the cutpoints $-10$, $-9$, ..., 0, ..., 9, 10. Comment on the shapes of the histograms relative to a $N(0, 1)$ density curve.

7. Suppose we have an urn containing 100 balls with 20 labeled 1, 50 labeled 2 and 30 labeled 3. Using sampling with replacement generate a sample of size 1000 from this distribution employing the **random** command to generate the sample directly from the relevant population distribution. Use the **table** command to record the proportion of each label in the sample.

# Chapter 4

# Probability: The Study of Randomness

In this chapter the concept of probability is introduced more formally than previously in the book. Probability theory underlies a powerful computational methodology known as simulation. Simulation has many applications in probability and statistics and also in many other fields such as engineering, chemistry, physics and economics. We discussed some aspects of this in Chapter 3 and we continue this here.

## 4.1  Basic Probability Calculations

The calculation of probabilities for random variables can often be simplified by tabulating the cumulative distribution function. Also means and variances are easily calculated using component-wise column operations in Minitab. For example, suppose we have the probability distribution

| $x$ | 1 | 2 | 3 | 4 |
|---|---|---|---|---|
| probability | .1 | .2 | .3 | .4 |

in columns C1 and C2 with the values in C1 and the probabilities in C2. Then the commands

```
MTB > parsums c2 c3
MTB > print c1 c3
 Row C1 C3
 1 1 0.1
 2 2 0.3
 3 3 0.6
 4 4 1.0
MTB > let c4=c1*c2
MTB > let c5=c1*c1*c2
MTB > let k1=sum(c4)
MTB > let k2=sum(c5)-k1*k1
MTB > print k1 k2
K1 3.00000
K2 1.00000
```

place the cumulative distribution function in C3, calculate the mean and variance and store these in K1 and K2 respectively. The mean is 3 and the variance is 1.

## 4.2   Simulation

The **random** command is the basic tool for carrying out simulations in Minitab. As we saw in II.3 the general syntax of the **random** command is

**random** V into $E_1 \ldots E_m$

and this puts a sample of size V into each of the columns $E_1$, ..., $E_m$ according to the distribution specified by the subcommand. If no subcommand is provided then this distribution is taken to be the $N(0, 1)$ distribution. To generate from the general $N(\mu, \sigma)$ we use the subcommand

**normal mu** $= V_1$ **sigma** $= V_2$

where $V_1$ is $\mu$ and $V_2$ is $\sigma$. There are many other distributions that can be sampled from. For example, the subcommand

**uniform** $V_1$ $V_2$

specifies the continuous uniform distribution on the interval $(V_1, V_2)$; i.e. subintervals of the same length have the same probability of occurring, while

**integer** $V_1$ $V_2$

specifies a uniform distribution on the integers from $V_1$ to $V_2$. If we have placed a discrete probability distribution in column $E_2$, on the values in column $E_1$, then the subcommand

**discrete** $E_1$ $E_2$

generates a sample from this distribution. The subcommand

**binomial** $V_1$ $V_2$

generates a sample from the *Binomial*$(n, p)$ distribution where $n$ is specified by $V_1$ and $p$ is specified by $V_2$. We introduce some other distributions in the exercises.

## 4.2.1 Simulation for Approximating Probabilities

Simulation can be used to approximate probabilities. For example, suppose we are asked to calculate

$$P(.1 \leq X_1 + X_2 \leq .3)$$

when $X_1, X_2$ are both independent and follow the uniform distribution on the interval $(0, 1)$. Then the commands

```
MTB > random 1000 c1 c2;
SUBC> uniform 0 1.
MTB > let c3=c1+c2
MTB > let c4 = .1<=c3 and c3<=.3
MTB > let k1=sum(c4)/n(c4)
MTB > print k1
K1 0.0400000
MTB > let k2=sqrt(k1*(1-k1)/n(c4))
MTB > print k2
K2 0.00619677
MTB > let k3=k1-3*k2
MTB > let k4=k1+3*k2
MTB > print k3 k4
K3 0.0214097
K4 0.0585903
```

generate $N = 1000$ independent values of $X_1, X_2$ and place these values in C1 and C2 respectively, then calculate the sum $X_1 + X_2$ and put these values in C3. Then using the comparison operators discussed in I.10.4, a 1 is recorded in C4 every time $.1 \leq X_1 + X_2 \leq .3$ is true and a 0 is recorded there otherwise. We then calculate the proportion of 1's in the sample as K1 and this is our estimate $\hat{p}$ of the probability. We will see later that a good measure of the accuracy of this estimate is the *standard error of the estimate* which in this case is given by

$$\sqrt{\hat{p}(1-\hat{p})/N}$$

and this is computed in K2. Actually we can feel fairly confident that the true value of the probability is in the interval

$$\hat{p} \pm 3\sqrt{\hat{p}(1-\hat{p})/N}$$

which in this case equals the interval $(0.0214097, 0.0585903)$. So we know the true value of the probability with reasonable accuracy. As the simulation size $N$ increases, the Law of Large Numbers says that $\hat{p}$ converges to the true value of the probability.

## 4.2.2   Simulation for Approximating Means

The means of distributions can also be approximated using simulations in Minitab. For example, suppose $X_1, X_2$ are both independent and follow the uniform distribution on the interval $(0, 1)$ and that we want to calculate the mean of $Y = 1/(1 + X_1 + X_2)$. Then we can approximate this in a simulation. The code

```
MTB > random 1000 c1 c2;
SUBC> uniform 0 1.
MTB > let c3=1/(1+c1+c2)
MTB > let k1=mean(c3)
MTB > let k2=stdev(c3)/sqrt(n(c3))
MTB > print k1 k2
K1 0.521532
K2 0.00375769
MTB > let k3=k1-3*k2
MTB > let k4=k1+3*k2
```

```
MTB > print k3 k4
K3 0.510259
K4 0.532805
```

generates $N = 1000$ independent values of $X_1, X_2$ and places these values in C1, C2, then calculates $Y = 1/(1 + X_1 + X_2)$ and puts these values in C3. The mean of C3 is stored in K1 and this is our estimate of the mean value of $Y$. As a measure of how accurate this estimate is we compute the standard error of the estimate which is given by the standard deviation divided by the square root of the simulation sample size $N$. Again we can feel fairly confident that the interval given by the estimate plus or minus 3 times the standard error of the estimate contains the true value of the mean. In this case this interval is given by $(0.510259, 0.532805)$ and so we know this mean with reasonable accuracy. As the simulation size $N$ increases, the Law of Large Numbers says that the approximation converges to the true value of the mean.

## 4.3  Exercises

When the data for an exercise come from an exercise in IPS, the IPS exercise number is given in parentheses ( ). All computations in these exercises are to be carried out using Minitab and the exercises are designed to ensure that you have a reasonable understanding of the Minitab material in this chapter. More generally you should be using Minitab to do all the computations and plotting required for the problems in IPS.

*If your version of Minitab places restrictions such that the value of the simulation sample size $N$ requested in these problems is not feasible, then substitute a more appropriate value. Be aware, however, that the accuracy of your results is dependent on how large $N$ is.*

1. Suppose we have the probability distribution

| $x$ | 1 | 2 | 3 | 4 | 5 |
|---|---|---|---|---|---|
| probability | .15 | .05 | .33 | .37 | .10 |

   on the values 1, 2, 3, 4 and 5. Using Minitab verify that this is a probability distribution. Make a bar chart (probability histogram) of this distribution. Tabulate the cumulative distribution. Calculate the mean and variance of this distribution. Suppose that three independent outcomes $(X_1, X_2, X_3)$ are generated from this distribution. Compute the probability that $1 < X_1 \leq 4, 2 \leq X_2$ and $3 < X_3 \leq 5$.

2. (4.26) Indicate how you would simulate the game of roulette using Minitab. Based on a simulation of $N = 1000$ estimate the probability of getting red and a multiple of 3. Also record the standard error of the estimate.

3. A probability distribution is placed on the integers 1, 2, ..., 100 where the probability of integer $i$ is $c/i^2$. Determine $c$ so that this is a probability distribution. What is the mean value? What is the $90th$ percentile? Generate a sample of 20 from the distribution.

4. The expression $e^{-x}$ for $x > 0$ is the density curve for what is called the *Exponential* (1) distribution. Plot this density curve in the interval from 0 to 10 using an increment of .1. The **random** command can be used to generate from this distribution using the subcommand **exponential** 1. Generate a sample of 1000 from this distribution and estimate its mean. Approximate the probability that a value generated from this distribution is in the interval (1,2). The general *Exponential* $(\lambda)$ has a density curve given by $\lambda^{-1}e^{-x/\lambda}$ for $x > 0$ and where $\lambda > 0$ is a fixed constant. Repeat the simulation with $\lambda = 3$. The general distribution can be generated from using the subcommand **exponential** V where V is equal to $\lambda$. Comment on the values of the estimated means.

5. Suppose you carry out a simulation to approximate the mean of a random variable $X$ and you report the value 1.23 with a standard error of .025. If you are then asked to approximate the mean of $Y = 3 + 5X$ do you have to carry out another simulation? If not, what is your approximation and what is the standard error of this approximation?

6. (4.50) Simulate 5 rounds of the game Keno where you bet on 10 each time. Calculate your total winnings (losses!).

7. Suppose that a random variable $X$ follows a $N(3, 2.3)$ distribution. Subsequently conditions change and no values smaller than $-1$ or bigger than 9.5 can occur; i.e. the distribution is conditioned to the interval

$(-1, 9.5)$. Generate a sample of 1000 from the truncated distribution and use the sample to approximate its mean.

8. Suppose that $X$ is a random variable and follows a $N(0, 1)$ distribution. Simulate $N = 1000$ values from the distribution of $Y = X^2$ and plot these values in a histogram with cutpoints 0, .5, 1, 1.5, ..., 15. Approximate the mean of this distribution. Now generate $Y$ directly from its distribution, which is known to be a $Chisquare(1)$ distribution. In general the $Chisquare(k)$ distribution can be generated from via the subcommand **chisquare** V where V is the *degrees of freedom* which in this case equals 1. Plot the $Y$ values in a histogram using the same cutpoints. Comment on the two histograms.

9. If $X_1$ and $X_2$ are independent random variables with $X_1$ following a $Chisquare(k_1)$ distribution and $X_2$ following a $Chisquare(k_2)$ distribution, then it is known that $Y = X_1 + X_2$ follows a $Chisquare(k_1 + k_2)$ distribution. For $k_1 = 1$, $k_2 = 1$ verify this empirically by plotting histograms with cutpoints 0, .5, 1, 1.5, ..., 15 based on simulations of size $N = 1000$.

10. If $X_1$ and $X_2$ are independent random variables with $X_1$ following a $N(0, 1)$ distribution and $X_2$ following a $Chisquare(k)$ distribution then it is known that

$$Y = \frac{X_1}{\sqrt{X_2/k}}$$

follows a $Student(k)$ distribution. The $Student(k)$ distribution can be generated from using the subcommand **student** V where V is the *degrees of freedom* which in this case equals $k$. For $k = 3$ verify this result empirically by plotting histograms with cutpoints $-10$, $-9$, ..., 9, 10, based on simulations of size $N = 1000$.

11 If $X_1$ and $X_2$ are independent random variables with $X_1$ following a $Chisquare(k_1)$ distribution and $X_2$ following a $Chisquare(k_2)$ distribution, then it is known that

$$Y = \frac{X_1/k_1}{X_2/k_2}$$

follows a $F(k_1, k_2)$ distribution. The $F(k_1, k_2)$ distribution can be generated from using the subcommand **F** $V_1$ $V_2$ where $V_1$ is the numerator degrees of freedom, equal to $k_1$, and $V_2$ is the denominator degrees of freedom equal to $k_2$. For $k_1 = 1$, $k_2 = 1$ verify this empirically by plotting histograms with cutpoints 0, .5, 1, 1.5, ..., 15 based on simulations of size $N = 1000$.

# Chapter 5

# From Probability to Inference

New Minitab commands discussed in this chapter
| | | | | |
|---|---|---|---|---|
| ichart | pchart | rchart | schart | xbarchart |

In this chapter the subject of statistical inference is introduced. Whereas we may feel fairly confident that the variation in a system can be described by probability it is typical that we don't know which probability distribution is appropriate. Statistical inference prescribes methods for using data derived from the contexts in question to choose appropriate probability distributions. For example, in a coin-tossing problem the *Bernoulli(p)* distribution is appropriate, when the tosses are independent, but what is an appropriate choice of $p$?

## 5.1 The Binomial Distribution

Suppose that $X_1, \ldots, X_n$ is a sample from the *Bernoulli(p)* distribution; i.e. $X_1, \ldots, X_n$ are independent realizations where each $X_i$ takes the value 1 or 0 with probabilities $p$ and $1 - p$ respectively. Then the random variable $Y = X_1 + \cdots + X_n$ equals the number of 1's in the sample and follows, as discussed in IPS, a *Binomial(n, p)* distribution. Therefore $Y$ can take on any of the values $0, 1, \ldots, n$ with positive probability. In fact an exact formula can be derived for these probabilities; namely

$$P(Y = k) = \binom{n}{k} p^k (1-p)^{n-k}$$

is the probability that $Y$ takes the value $k$ for $0 \leq k \leq n$. When $n$ and $k$ are small then this formula could be used to evaluate this probability but it is almost always better to use software like Minitab to do it and when these values are not small, it is necessary. Also we can use Minitab to compute the $Binomial(n, p)$ cumulative probability distribution –. the probability contents of intervals $(-\infty, x]$ – and the inverse cumulative distribution – percentiles of the distribution.

For individual probabilities we use the **pdf** command with the **binomial** subcommand. For example, suppose we have a $Binomial(30, .2)$ distribution and want to compute the probability $P(Y = 10)$. Then the command

```
MTB > pdf 10;
SUBC> binomial 30 .2.
Binomial with n = 30 and p = 0.200000
     x      P( X = x)
   10.00       0.0355
```

computes this probability as .0355. If we want to compute the probability of getting 10 or fewer successes, then this is the probability of the interval $(-\infty, 10]$ and we can use the **cdf** command

```
MTB > cdf 10;
SUBC> binomial 30 .2.
Binomial with n = 30 and p = 0.200000
     x      P( X <= x)
   10.00       0.9744
```

which computes this probability as .9744. Suppose we want to compute the first quartile of this distribution. Then the **invcdf** command

```
MTB > invcdf .25;
SUBC> binomial 30 .2.
Binomial with n = 30 and p = 0.200000
   x     P( X <= x)      x      P( X <= x)
   3       0.1227        4        0.2552
```

gives the values $x$ that have cumulative probabilities just smaller and just larger than the value requested. Recall that with a discrete distribution, such as the $Binomial(n, p)$, we will not in general be able to obtain an exact percentile.

The general syntax of the **pdf**, **cdf** and **invcdf** commands is given in II.1.4 and recall that these commands can operate on all the values in a column simultaneously. Actually when $n$ is large even software will not suffice and you will have to use the normal approximation as discussed in IPS. Of course the **pdf** and **cdf** commands with the **normal** subcommand can be used for this.

We might also want to simulate from the $Binomial(n, p)$ distribution. For this we use the **random** command with the **binomial** subcommand. For example,

```
MTB > random 10 c1;
SUBC> binomial 30 .2.
MTB > print c1
C1
 2 2 4 2 11 5 7 8 5 2
```

generates a sample of 10 from the $Binomial(30, .2)$ distribution.

## 5.2   Control Charts

Control charts are used to monitor a process to ensure that it is under statistical control. There is a wide variety of such charts depending on the statistic used for the monitoring and the test used to detect when a process is out of control.

## 5.3   The XBARCHART Command

Suppose we have placed a random sample of 100 from the $N(5, 2)$ distribution in C1. Then the command

```
MTB > xbarchart c1 5;
SUBC> mu=5;
SUBC> sigma=2.
```

produces a character $\bar{x}$ chart like the high resolution plot given in Figure 5.1. The 100 observations are partitioned into successive groups of 5 and $\bar{x}$ is plotted for each of these. The center line of the chart is at the mean 5 and

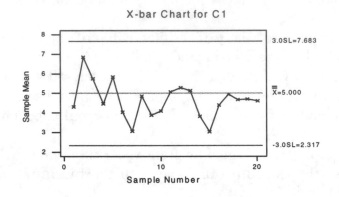

Figure 5.1: An $\bar{x}$-chart for a random sample of 100 from the $N(5,2)$ distribution using historical values for $\mu = 5$ and $\sigma = 2$.

the lines 3 standard deviations above and below the center line are drawn at $5 + 3 \cdot 2/\sqrt{5} = 7.683$ and at $5 - 3 \cdot 2/\sqrt{5} = 2.317$ respectively. The chart confirms that these choices for $\mu$ and $\sigma$ are reasonable as we might expect.

Of course we will typically not know the true values of $\mu$ and $\sigma$ and these must be estimated from the data. The command

```
MTB > xbarchart c1 5
```

draws the plot given in Figure 5.2. Here $\mu$ is estimated by the overall mean and $\sigma$ is estimated by pooling the sample deviations for each subgroup. Notice that using the test of one mean above the control limit this chart indicates a loss of control.

The syntax of the **xbarchart** command is

**xbarchart** $E_1$ $E_2$

where $E_1$ is a column containing the data and $E_2$ is either a constant, indicating how many observations are used to define a subgroup, or a column of values, indicating how the elements of $E_1$ are to be grouped for the calculation of the means. When $E_2$ equals 1, $\sigma$ cannot be estimated using standard deviations and an alternative estimator is used.

There are various subcommands that can be used with **xbarchart.** For example, we have already seen how **mu** and **sigma** can be used to specify the population mean and standard deviation. Sometimes the data may come

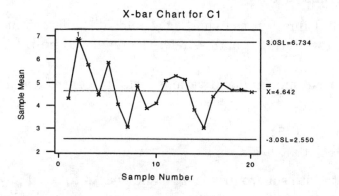

Figure 5.2: An $\bar{x}$-chart for a random sample of 100 from the $N(5,2)$ distribution using estimated values for $\mu$ and $\sigma$.

with each subgroup in a separate row. Then the **rsub** subcommand can be used provided the subgroups are of equal size. For example, the code

```
MTB > set c2
DATA> 20(1:5)
DATA> end
MTB > unstack c1 c3-c7;
SUBC> subscripts c2.
MTB > xbar;
SUBC> rsub c3 c4 c5 c6 c7.
```

breaks the observations in C1 into subgroups of 5 with the first subgroup in the first row of C3-C7, etc. The default upper and lower control limits are at three standard deviations from the mean. The **slimits** subcommand allows you to control the multiples of the standard deviation for which you want control limits drawn. For example,

```
MTB > xbar c1 5;
SUBC> slimits 1 2 6.
```

draws control limits one, two and six standard deviations on either side of the center line. The **hlines** subcommand allows you to draw lines on the chart at fixed ordinates. For example,

```
MTB > xbar c1 5;
```

```
SUBC> hlines 4 5.
```

draws horizantal lines on the chart at 4 and 5. Different methods can be prescribed for estimating $\sigma$ and we refer the reader to **help** for a description of these.

Using the **test** subcommand various tests for control can be carried out. For example,

```
MTB > xbar c1 5;
SUBC> test 1.
```

breaks the data into subgroups of size 5 and checks to see if any of the points are outside the control limits. The subcommand **test** 2 checks to see if there are 9 points in a row on the same side of the center line, **test** 3 checks to see if there are 6 points in a row all increasing or all decreasing. There are a total of 8 tests like this all looking for patterns. The subcommand **test 1:8** performs all 8 tests.

In the window environment $\bar{x}$ charts can also be carried out using the menu command S̲tat ▶ Control C̲harts ▶ X̲bar and filling in the dialog box appropriately. The session commands work a little differently than we have just described when obtaining high resolution plots and we recommend using the menu commands for these as we do for all high resolution plots.

## 5.4   Other Control Charts

There are many other control charts available within Minitab. For example, the command **pchart** produces $p$-charts. A $p$-chart is appropriate when a response is coming from a *Binomial* $(n, p)$ distribution, e.g. the count of the number of defectives in a batch of size $n$, and we use the proportion of defectives $\hat{p}$ to control the process. For example, suppose that C1 contains a sample of 100 from the *Binomial*(.4) distribution. Then the command

```
MTB > pchart c1 5
```

produces a plot like the high resolution plot shown in Figure 5.3 where the proportion of defectives in each group of 5 is plotted against time. The asymmetry of the plot is caused by the fact that $\hat{p}$ can never be smaller than 0. Also rather than plotting subgroup means we can plot subgroup standard deviations using the command **schart** and subgroup ranges using **rchart**. An individual chart plot is simply an $\bar{x}$ chart with the number of

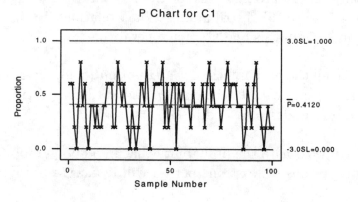

Figure 5.3: A *P*-chart for 100 randomly generated values from the *Binomial*(5, .4) distribution.

observations equal to 1 and is accessed via the command **ichart**. All of these and other control charts available in Minitab have a set of subcommands similar to **xbarchart**.

In the window environment these and other plots can be accessed via Stat ▶ Control Charts and clicking on the relevant chart in the list. You also have the ability to design tests of control using Stat ▶ Control Charts ▶ Define Tests and filling in the dialog box.

# 5.5 Exercises

When the data for an exercise come from an exercise in IPS, the IPS exercise number is given in parentheses ( ). All computations in these exercises are to be carried out using Minitab and the exercises are designed to ensure that you have a reasonable understanding of the Minitab material in this chapter. More generally you should be using Minitab to do all the computations and plotting required for the problems in IPS.

*If your version of Minitab places restrictions such that the value of the simulation sample size N requested in these problems is not feasible, then substitute a more appropriate value. Be aware, however, that the accuracy of your results is dependent on how large N is.*

1. Calculate all the probabilities for the *Binomial*(5, .4) distribution and the *Binomial*(5, .6) distribution. What relationship do you observe? Can you explain this and state a general rule?

2. Compute all the probabilities for a *Binomial*(5, .8) distribution and use these to directly calculate the mean and variance. Verify your answers using the formulas provided in IPS.

3. (5.17) Approximate the probability that in 50 polygraph tests given to truthful persons at least one person will be accused of lying.

4. Generate $N = 1000$ samples of size $n = 5$ from the $N(0, 1)$ distribution. Record a histogram for $\bar{x}$ using the cutpoints $-3, -2.5, -2, ...,$ 2.5, 3.0. Generate a sample of size $N = 1000$ from the $N(0, 1/\sqrt{5})$ distribution. Plot the histogram using the same cutpoints and compare the histograms. What will happen to these histograms as we increase $N$?

5. Generate $N = 1000$ values of $X_1, X_2$ where $X_1$ follows a $N(3, 2)$ distribution and $X_2$ follows a $N(-1, 3)$ distribution. Compute $Y = X_1 - 2X_2$ for each of these pairs and plot a histogram for $Y$ using the cutpoints $-20, -15, ..., 25, 30$. Generate a sample of $N = 1000$ from the appropriate distribution of $Y$ and plot a histogram using the same cutpoints.

6. Plot the density curve for the *Exponential*(3) distribution (see Exercise II.4.4) between 0 and 15 with an increment of .1. Generate $N = 1000$ samples of size $n = 2$ from the *Exponential*(3) distribution and record the sample means. Standardize the sample of $\bar{x}$ using $\mu = 3$ and $\sigma = 3$. Plot a histogram of the standardized values using the cutpoints $-5$, $-4, ..., 4, 5$. Repeat this for $n = 5, 10$. Comment on the shapes of these histograms. See Example 5.18 in IPS for further discussion of this distribution.

7. Plot the density of the uniform distribution on (0,1). Generate $N = 1000$ samples of size $n = 2$ from this distribution. Standardize the sample of $\bar{x}$ using $\mu = .5$ and $\sigma = \sqrt{1/12}$. Plot a histogram of the standardized values using the cutpoints $-5, -4, ..., 4, 5$. Repeat this for $n = 5, 10$. Comment on the shapes of these histograms.

8. The *Weibull* $(\beta)$ has density curve $\beta x^{\beta-1} e^{-x^\beta}$ for $x > 0$ where $\beta > 0$ is a fixed constant. Plot the *Weibull* $(2)$ density in the range 0 to 10 with an increment of .1. See section 5.2 in IPS for discussion of this distribution. Generate a sample of $N = 1000$ from this dustribution using the subcommand **weibull** $V_1$ $V_2$ where $V_1$ is the value of $\beta$ and $V_2$ is the value of a general scaling parameter which in the case of the *Weibull* $(\beta)$ distribution equals 1. Plot a probability histogram and compare with the density curve.

9. (5.50) Make an $\bar{x}$-chart for this data with three sigma control lines using **xbarchart** and its subcommands. What tests for control are failed?

10. (5.59) Make a $p$-chart for this data with three sigma control lines using **xbarchart** and its subcommands. What tests for control are failed?

# Chapter 6

# Introduction to Inference

**New Minitab commands discussed in this chapter**

**zinterval**   **ztest**

In this chapter the basic tools of statistical inference are discussed. There are a number of Minitab commands that aid in the computation of confidence intervals and for carrying out tests of significance.

## 6.1   The ZINTERVAL Command

The **zinterval** command computes confidence intervals for the mean $\mu$ using a sample $x_1, \ldots, x_n$ from a distribution where we know the standard deviation $\sigma$. There are three situations when this is appropriate:

(1) We know that we are sampling from a normal distribution with unknown mean $\mu$ and known standard deviation $\sigma$ and thus

$$z = \frac{\bar{x} - \mu}{\sigma/\sqrt{n}}$$

is distributed $N(0,1)$.

(2) We have a large sample from a distribution with unknown mean $\mu$ and known standard deviation $\sigma$ and the central limit theorem approximation to

111

the distribution of $\bar{x}$ is appropriate; i.e. the distribution of

$$z = \frac{\bar{x} - \mu}{\sigma/\sqrt{n}}$$

is approximately distributed $N(0,1)$.

(3) We have a large sample from a distribution with unknown mean $\mu$ and unknown standard deviation $\sigma$ and the sample size is large enough so that

$$z = \frac{\bar{x} - \mu}{s/\sqrt{n}}$$

is approximately $N(0,1)$ where $s$ is the sample standard deviation.

The confidence interval takes the form $\bar{x} \pm z^*\sigma/\sqrt{n}$, where $s$ is substituted for $\sigma$ in case (3), and $z^*$ is determined from the $N(0,1)$ distribution by the confidence level desired, as described in IPS. Of course situation (3) is probably the most realistic but note that the confidence intervals constructed for (1) are exact while those constructed under (2) and (3) are only approximate and a larger sample size is required in (3) for the approximation to be reasonable than for (2).

Consider Example 6.2 in IPS and suppose the data 190.5, 189.0, 195.5, 187.0 are stored in C1 and, as mentioned in the text, it makes sense to take $\sigma = 3$. Then the command

```
MTB > zinterval 90 sigma=3 C1
The assumed sigma = 3.00
Variable N    Mean StDev SE Mean      90.0 % CI
C1        4 190.50  3.63    1.50 ( 188.03, 192.97)
```

produces the 90% confidence interval (188.03, 192.97) for $\mu$. The general syntax of the **zinterval** command is

**zinterval** $V_1$ sigma $= V_2$ $E_1 \ldots E_m$

where $V_1$ is the confidence level and is any value between 1 and 99.99, $V_2$ is the assumed value of $\sigma$ and $E_1$, ..., $E_m$ are columns of data. A $V_1\%$ confidence interval is produced for each column specified. If no value is specified for $V_1$ then the default value is 95%.

In the window environment these confidence intervals are produced via the command S̲tat ▶ B̲asic Statistics ▶ 1- Sample Z̲ and filling in the dialog box appropriately.

## 6.2   The ZTEST Command

The **ztest** command is used when we want to test the hypothesis that the unknown mean $\mu$ equals a value $\mu_0$ and one of the situations (1), (2) or (3) obtains as discussed in II.6.1. The test is based on computing a $P$-value using the observed value of

$$z = \frac{\bar{x} - \mu_0}{\sigma/\sqrt{n}}$$

and the $N(0,1)$ distribution as described in IPS.

Consider Example 6.6 in IPS where we are asked to test the null hypothesis $H_0 : \mu = 187$ against the alternative $H_a : \mu > 187$ and suppose the data 190.5, 189.0, 195.5, 187.0 are stored in C1 and, as mentioned in the text, it makes sense to take $\sigma = 3$. The command

```
MTB > ztest 187 sigma=3 c1;
SUBC> alternative 1.
The assumed sigma = 3.00
Variable  N    Mean StDev SE Mean    Z       P
C1        4 190.50  3.63    1.50  2.33  0.0099
```

gives the $P$-value for this test as .0099.

The general syntax of the **ztest** command is

**ztest** $V_1$ sigma $= V_2\ E_1 \ldots E_m$

where $V_1$ is the hypothesized value to be tested, $V_2$ is the assumed value of $\sigma$ and $E_1$, ..., $E_m$ are columns of data. If no value is specified for $V_1$, then the default is 0. A test of the hypothesis is carried out for each column. If no **alternative** subcommand is specified, then a two-sided test is conducted, i.e. $H_0 : \mu = V_1$ against the alternative $H_a : \mu \neq V_1$. If the subcommand

```
SUBC> alternative 1.
```

is used, then a test of $H_0 : \mu = V_1$ against the alternative $H_a : \mu > V_1$ is conducted. If the subcommand

```
SUBC> alternative -1.
```

is used, then a test of $H_0 : \mu = V_1$ against the alternative $H_a : \mu < V_1$ is conducted.

In the window environment these tests are produced via the command Stat ▶ Basic Statistics ▶ 1- Sample Z and filling in the dialog box appropriately.

## 6.3   Simulations for Confidence Intervals

When we are sampling from a $N(\mu, \sigma)$ distribution and know the value of $\sigma$ then the confidence intervals constructed in II.6.1 using **zinterval** are exact; i.e. in the long run a proportion 95% of the 95% confidence intervals constructed for an unknown mean $\mu$ will contain the true value of this quantity. Of course any given confidence interval may or may not contain the true value of $\mu$ and in any finite number of such intervals so constructed some proportion other than 95% will contain the true value of $\mu$. As the number of intervals increases, however, the proportion covering will go to 95%.

We illustrate this via a simulation study based on computing 90% confidence intervals. The commands

```
MTB > random 100 c1-c5;
SUBC> normal 1 2.
MTB > rmean c1-c5 c6
MTB > invcdf .95;
SUBC> normal 0 1.
Normal with mean = 0 and standard deviation = 1.00000
 P( X <= x) x
 0.9500 1.6449
MTB > let k1=1.6449*2/sqrt(5)
MTB > let c7=c6-k1
MTB > let c8=c6+k1
MTB > let c9=c7<1 and c8>1
MTB > mean c9
 Mean of C9 = 0.94000
MTB > set c10
DATA> 1:25
DATA> end
MTB > delete 26:100 c7 c8
MTB > mplot c7 versus c10 c8 versus c10;
SUBC> xstart=1 end=25;
SUBC> xincrement=1.
```

generate 100 random samples of size 5 from the $N(1,2)$ distribution, place the means in C6, the lower end-point of a 90% confidence interval in C7 and the upper end-point in C8 and then record whether or not a confidence interval covers the true value $\mu = 1$ by placing a 1 or 0 in C9 respectively.

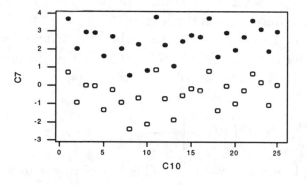

Figure 6.1: Plot of 90% confidence intervals for the mean when sampling from the $N(1, 2)$ distribution with $n = 5$. The lower end-point is open and the upper end-point is closed.

The mean of C9 is the proportion of intervals that cover and this is 94% which is 4% too high. Finally we plotted the first 25 of these intervals in a plot much like the high resolution plot given in Figure 6.1. Drawing a solid horizontal line at 1 on the $y$-axis indicates that most of these intervals do indeed cover the true value $\mu = 1$.

The simulation just carried out simply verifies a theoretical fact. On the other hand when we are computing approximate confidence intervals – i.e. we are not sampling necessarily from a normal distribution – it is good to do some simulations from various distributions to see how much reliance we can place in the approximation at a given sample size. The true *coverage probability* of the interval, i.e. the long-run proportion of times that the interval covers the true mean, will not in general be equal to the nominal confidence level. Small deviations are not serious but large ones are.

## 6.4 Simulations for Power Calculations

It is also useful to know in a given context how sensitive a particular test of significance is. By this we mean how likely it is that the test will lead us to reject the null hypothesis when the null hypothesis is false. This is measured by the concept of the *power* of a test. Typically a level $\alpha$ is chosen for the

$P$-value at which we would definitely reject the null hypothesis if the $P$-value is smaller than $\alpha$. For example, $\alpha = .05$ is a common choice for this level. Suppose then that we have chosen the level of $.05$ for the two-sided $z$-test and we want to evaluate the power of the test when the true value of the mean is $\mu = \mu_1$, i.e. evaluate the probability of getting a $P$-value smaller than $.05$ when the mean is $\mu_1$. The two-sided $z$-test with level $\alpha$ rejects $H_0 : \mu = \mu_0$ whenever

$$P\left(|Z| > \left|\frac{\bar{x} - \mu_0}{\sigma/\sqrt{n}}\right|\right) \leq \alpha$$

where $Z$ is a $N(0,1)$ random variable. This is equivalent to saying that the null hypothesis is rejected whenever

$$\left|\frac{\bar{x} - \mu_0}{\sigma/\sqrt{n}}\right|$$

is greater than or equal to the $1 - \alpha/2$ percentile for the $N(0,1)$ distribution. For example, if $\alpha = .05$ then $1 - \alpha/2 = .975$ and this percentile can be obtained via the command

```
MTB > invcdf .975;
SUBC> normal 0 1.
Normal with mean = 0 and standard deviation = 1.00000
   P( X <= x)          x
     0.9750        1.9600
```

which gives the value $1.96$. Denote this percentile by $z^*$. Now if $\mu = \mu_1$ then

$$\frac{\bar{x} - \mu_0}{\sigma/\sqrt{n}}$$

is a realized value from the distribution of $Y = \frac{\bar{X} - \mu_0}{\sigma/\sqrt{n}}$ when $\bar{X}$ is distributed $N(\mu_1, \sigma/\sqrt{n})$. Therefore $Y$ follows a $N(\frac{\mu_1 - \mu_0}{\sigma/\sqrt{n}}, 1)$ distribution. Then the power of the two-sided test at $\mu = \mu_1$ is

$$P(|Y| > z^*)$$

and this can be evaluated exactly using the **cdf** command with the **normal** subcommand after writing

$$
\begin{aligned}
P(|Y| > z^*) &= P(Y > z^*) + P(Y < -z^*) \\
&= P(Z > (\mu_1 - \mu_0) + \frac{\sigma}{\sqrt{n}}z^*) + P(Z < (\mu_1 - \mu_0) - \frac{\sigma}{\sqrt{n}}z^*)
\end{aligned}
$$

with $Z$ following a $N(0,1)$ distribution.

This derivation of the power of the two-sided test depended on the sample coming from a normal distribution as this leads to $\bar{X}$ having an exact normal distribution. In general, however, $\bar{X}$ will only be approximately normal and so the normal calculation is not exact. To assess the effect of the nonnormality, however, we can often simulate sampling from a variety of distributions and estimate the probability $P(|Y| > z^*)$. For example, suppose that we want to test $H_0 : \mu = 0$ in a two-sided $z$-test based on a sample of 10, where we estimate $\sigma$ by the sample standard deviation and we want to evaluate the power at 1. Let us further suppose that we are actually sampling from a uniform distribution on the interval $(-10, 12)$ which indeed has its mean at 1. Then the simulation

```
MTB > random 1000 c1-c10;
SUBC> uniform -10 12.
MTB > rmean c1-c10 c11
MTB > rstdev c1-c10 c12
MTB > let c13=absolute(c11/(c12/sqrt(10)))
MTB > let c14=c13>1.96
MTB > let k1=mean(c14)
MTB > let k2=sqrt(k1*(1-k1)/n(c14))
MTB > print k1 k2
K1 0.112000
K2 0.00997276
```

estimates the power to be .112 and the standard error of this estimate, as given in K2, is approximately .01. The application determines whether or not the assumption of a uniform distribution makes sense and whether or not this power is indicative of a sensitive test or not.

In the window environment exact power calculations can be carried out under the assumption of sampling from a normal distribution using Power and Sample Size ▶ 1-Sample Z and filling in the dialog box appropriately. Also the minimum sample size required to guarantee a given power at a prescribed difference $|\mu_1 - \mu_0|$ can be obtained using this command.

## 6.5   The Chi-square Distribution

If $Z$ is distributed according to the $N(0,1)$ distribution then $Y = Z^2$ is distributed according to the $Chisquare(1)$ distribution. If $X_1$ is distributed $Chisquare(k_1)$ independent of $X_2$ distributed $Chisquare(k_2)$ then $Y = X_1 + X_2$ is distributed according to the $Chisquare(k_1 + k_2)$ distribution. There are Minitab commands that assist in carrying out computations for the $Chisquare(k)$ distribution. Note that $k$ is any nonnegative value and is referred to as the *degrees of freedom*.

The values of the density curve for the $Chisquare(k)$ distribution can be obtained using the **pdf** command with the **chisquare** subcommand. For example, the command

```
MTB > pdf c1 c2;
SUBC> chisquare 4.
```

calculates the value of the $Chisquare(4)$ density curve at each value in C1 and stores these values in C2. This is useful for plotting the density curve. The **cdf**, **invcdf** and **random** commands can also be used with the **chisquare** subcommand in this way to obtain values of the $Chisquare(k)$ cumulative distribution function and inverse distribution function and to obtain random samples from the distribution respectively.

We will see applications of the chi-square distribution later in the book but we mention one here. In particular, if $x_1, \ldots, x_n$ is a sample from a $N(\mu, \sigma)$ distribution then $(n-1)\, s^2/\sigma^2 = \sum_{i=1}^{n} (x_i - \bar{x})^2 / \sigma^2$ is known to follow a $Chisquare(n-1)$ distribution and this fact is used as a basis for inference about $\sigma$ (confidence intervals and tests of significance). Because of the nonrobustness of these inferences to small deviations from normality these inferences are not recommended.

In the window environment the density curve, cumulative distribution function and inverse distribution function can be evaluated via Calc ▶ Probability Distributions ▶Chi-Square while random samples are obtained using Calc ▶ Random Data ▶ Chi-Square.

## 6.6   Exercises

When the data for an exercise come from an exercise in IPS, the IPS exercise number is given in parentheses ( ). All computations in these exercises are to be carried out using Minitab and the exercises are designed to ensure that you have a reasonable understanding of the Minitab material in this chapter. More generally you should be using Minitab to do all the computations and plotting required for the problems in IPS.

*If your version of Minitab places restrictions such that the value of the simulation sample size N requested in these problems is not feasible, then substitute a more appropriate value. Be aware, however, that the accuracy of your results is dependent on how large N is.*

1. (6.9) Use the **zinterval** command to compute 90%, 95% and 99% confidence intervals for $\mu$.

2. (6.39) Use the **ztest** command to test the null hypothesis against the appropriate alternative. Evaluate the power of the test with level $\alpha = .05$ at $\mu = 33$.

3. Simulate $N = 1000$ samples of size 5 from the $N(1, 2)$ distribution and calculate the proportion of .90 confidence intervals for the mean that cover the true value $\mu = 1$.

4. Simulate $N = 1000$ samples of size 10 from the uniform distribution on (0,1) and calculate the proportion of .90 confidence intervals for the mean that cover the true value $\mu = .5$. Use $\sigma = 1/\sqrt{12}$.

5. .Simulate $N = 1000$ samples of size 10 from the *Exponential*(1) distribution (see Exercise II.4.4) and calculate the proportion of .95 confidence intervals for the mean that cover the true value $\mu = 1$. Use $\sigma = 3$.

6. The density curve for the *Student*(1) distribution takes the form

$$\frac{1}{\pi} \frac{1}{1 + x^2}$$

for $-\infty < x < \infty$. This special case is called the *Cauchy* distribution. Plot this density curve in the range $(-20, 20)$ using an increment of .1. Simulate $N = 1000$ samples of size 5 from the *Student*(1) distribution (see Exercise II.4.10) and calculate the proportion of .90 confidence intervals for the mean, using the sample standard deviation for $\sigma$, that cover the value $\mu = 0$. It is possible to obtain very bad approximations in this example because the central limit theorem does not apply to this distribution. In fact it does not have a mean.

7. The uniform distribution on the interval $(a, b)$ has mean $\mu = (a + b)/2$ and standard deviation $\sigma = \sqrt{(b - a)^2/12}$. Calculate the power at $\mu = 1$ of the two-sided $z$-test at level $\alpha = .95$ for testing $H_0 : \mu = 0$ when the sample size is $n = 10$, $\sigma$ is the standard deviation of a uniform distribution on $(-10, 12)$ and we are sampling from a normal distribution. Compare your result with the example in II.6.4.

8. Suppose that we are testing $H_0 : \mu = 0$ in a two-sided test based on a sample of 3. Approximate the power of the $z$-test at level $\alpha = .1$ at $\mu = 5$ when we are sampling from the distribution of $Y = 5 + W$ where $W$ follows a *Student*(6) distribution (see Exercise II.4.10) and we use the sample standard deviation to estimate $\sigma$. Note that the mean of the distribution of $Y$ is 5.

# Chapter 7

# Inference for Distributions

**New Minitab commands discussed in this chapter**

sinterval   stest   tinterval   ttest   twosample   twot

## 7.1   The Student Distribution

If $Z$ is distributed $N(0,1)$ independent of $X$ distributed $Chisquare(k)$ (see II.6.5), then

$$T = \frac{Z}{\sqrt{X/k}}$$

is distributed according to the $Student(k)$ distribution. The value $k$ is referred to as the *degrees of freedom* of the Student distribution. There are Minitab commands that assist in carrying out computations for this distribution.

The values of the density curve for the $Student(k)$ distribution can be obtained using the **pdf** command with the **student** subcommand. For example, the command

```
MTB > pdf c1 c2;
SUBC> student 4.
```

calculates the value of the $Student(4)$ density curve at each value in C1 and stores these values in C2. This is useful for plotting the density curve. The **cdf**, **invcdf** and **random** commands can also be used with the **student** subcommand in this way to obtain values of the $Student(k)$ cumulative distribution function and inverse distribution function and to obtain random samples from the distribution respectively.

In the window environment the density curve, cumulative distribution function and inverse distribution function can be evaluated via C̲alc ▶ Prob̲ability D̲istributions ▶ t̲ while random samples are obtained using C̲alc ▶ R̲andom Data ▶ t̲.

## 7.2   The TINTERVAL Command

When sampling from the $N(\mu, \sigma)$ distribution with $\mu$ and $\sigma$ unknown, an exact $1 - \alpha$ confidence interval for $\mu$ based on the sample $x_1, \ldots, x_n$ is given by $\bar{x} \pm t^* s/\sqrt{n}$ where $t^*$ is the $1 - \alpha/2$ percentile of the $Student(n - 1)$ distribution. These intervals can be obtained using the **tinterval** command. Suppose we place the data in Example 7.1 of IPS in column C1. Then the command

```
MTB > tinterval .95 c1
Variable    N    Mean  StDev  SE Mean      95.0 % CI
C1          8   22.50   7.19     2.54  ( 16.48, 28.52)
```

computes a 95% confidence interval for $\mu$ as (16.48, 28.52).

Note that the **tinterval** command can also be used to obtain confidence intervals for the difference of two means in a *matched pairs* design. For this store the difference of the measurements in a column and apply **tinterval** to that column.

The general syntax of the **tinterval** command is

**tinterval** V $E_1 \ldots E_m$

where V is the confidence level and is any value between 1 and 99.99 and $E_1$, ..., $E_m$ are columns of data. A V% confidence interval is produced for each column specified. If no value is specified for V then the default value is 95%.

In the window environment $t$ confidence intervals can be obtained using S̲tat ▶ B̲asic Statistics ▶ 1̲-Sample t and filling in the dialog box appropriately.

# 7.3 The TTEST Command

The **ttest** command is used when we we have a sample $x_1, \ldots, x_n$ from a normal distribution with unknown mean $\mu$ and standard deviation $\sigma$ and we want to test the hypothesis that the unknown mean equals a value $\mu_0$. The test is based on computing a $P$-value using the observed value of

$$t = \frac{\bar{x} - \mu_0}{s/\sqrt{n}}$$

and the $Student(n-1)$ distribution as described in IPS.

Suppose we have the data of Example 7.2 stored in column C1 and we want to test the null hypothesis $H_0 : \mu = 40$ versus the alternative $H_a : \mu \neq 40$. The command

```
MTB > ttest 40 c1
Test of mu = 40.00 vs mu not = 40.00
Variable    N     Mean    StDev   SE Mean      T        P
C1          8    22.50    7.19      2.54    -6.88   0.0002
```

calculates the $P$-value as .0002.

The general syntax of the **ttest** command is

**ttest** V $E_1 \ldots E_m$

where V is the hypothesized value to be tested and $E_1, \ldots, E_m$ are columns of data. If no value is specified for V, then the default is 0. A test of the hypothesis is carried out for each column. Also the **alternative** subcommand is available and works just as with the **ztest** command.

Note that the **ttest** command can also be used to carry out $t$ tests for the difference of two means in a matched pairs design. For this store the difference of the measurements in a column and apply **ttest** to that column.

We can calculate the power of the $t$ test using simulations. Note, however, that we must prescribe not only the mean $\mu_1$ but the standard deviation $\sigma_1$ as well, as there are two unknown parameters. For example,

```
MTB > random 1000 c1-c5;
SUBC> normal 2 3.
MTB > rmean c1-c5 c6
MTB > rstdev c1-c5 c7
MTB > let c8= absolute(c6/(c7/sqrt(5)))
MTB > invcdf .975;
```

```
SUBC> student 4.
Student's t distribution with 4 DF
 P( X <= x)          x
     0.9750       2.7764
MTB > let c9=c8>2.7764
MTB > let k1=mean(c9)
MTB > let k2=sqrt(k1*(1-k1)/n(c9))
MTB > print k1 k2
K1 0.194000
K2 0.0125046
```

carries out a simulation to approximate the power of the $t$ test for testing $H_0 : \mu = 0$ versus the alternative $H_a : \mu \neq 0$ at level $\alpha = .05$ and with $\mu_1 = 2$, $\sigma_1 = 3$. The power is estimated to be .194 with standard error .013. There is a more efficient way to carry out this power simulation using the noncentral Student distribution. This is discussed in Exercises II.7.9 and II.7.10.

In the window environment $t$ tests can be carried out using $\underline{S}$tat ▶ $\underline{B}$asic Statistics ▶ $\underline{1}$-Sample t and filling in the dialog box appropriately. Also in the window environment exact power calculations can be carried under the assumption of sampling from a normal distribution using $\underline{P}$ower and Sample Size ▶ 1-Sample t and filling in the dialog box appropriately. The exact power for the above problem is computed to be .2113 and is in agreement with our sampling results when sampling error is taken into account. Further the minimum sample size required to guarantee a given power at a prescribed difference $|\mu_1 - \mu_0|$ and standard deviation $\sigma$ can be obtained using this command.

## 7.4   The SINTERVAL and STEST Commands

As discussed in IPS, sometimes we cannot sensibly assume normality or transform to normality or make use of large samples so that there is a central limit theorem effect. In such a case we attempt to use *distribution free* or *nonparametric* methods. The methods based on the *sign test statistic* for the median is one of these.

For example, suppose we have the data for Example 7.1 in IPS stored in column C1. Then the **sinterval** command

```
MTB > sinterval 95 c1
```

```
Sign confidence interval for median
                    Achieved
        N    Median Confidence     Confidence interval Position
   C1   8    22.50   0.9297           ( 14.00, 31.00) 2
                     0.9500           ( 13.81, 31.00) NLI
                     0.9922           ( 11.00, 31.00) 1
```

computes a 95% confidence interval for the median. As the distribution of the sign statistic is discrete, in general the exact confidence cannot be attained so Minitab records the confidence intervals with confidence level just smaller and just greater than the confidence level requested and then records a middle interval by interpolation. The **stest** command in

```
MTB > stest 40 c1
Sign test of median = 40.00 versus not = 40.00
        N    Below   Equal    Above      P     Median
   C1   8      8       0        0     0.0078    22.50
```

computes the $P$-value for testing that the median of the population distribution equals 40 against the alternative that it doesn't. In this case the $P$-value is .0078. Also the sample median of 22.5 is recorded.

The general syntax of the **sinterval** command is

**sinterval** V $E_1 \ldots E_m$

where V is the confidence level and is any value between 1 and 99.99 and $E_1$, ..., $E_m$ are columns of data. A V% confidence interval is produced for each column specified. If no value is specified for V, then the default value is 95%.

The general syntax of the **stest** command is

**stest** V $E_1 \ldots E_m$

where V is the hypothesized value to be tested and $E_1$, ..., $E_m$ are columns of data. If no value is specified for V, then the default is 0. A test of the hypothesis is carried out for each column. The **alternative** subcommand is also available for one-sided tests.

Note that the **sinterval** and **stest** commands can also be used to construct confidence intervals and carry out tests for the median of a difference in a matched pairs design. For this store the difference of the measurements in a column and apply **sinterval** and **stest** to that column.

In the window environment these procedures can be carried out using $\underline{S}$tat ▶ $\underline{N}$onparametrics ▶ $\underline{1}$-Sample Sign and filling in the dialog box appropriately.

## 7.5   The TWOSAMPLE Command

If we have independent samples $x_{11}, \ldots x_{1n_1}$ from the $N(\mu_1, \sigma_1)$ distribution and $x_{12}, \ldots x_{1n_2}$ from the $N(\mu_2, \sigma_2)$ distribution where $\sigma_1$ and $\sigma_2$ are known then we can base inferences about the difference of the means $\mu_1 - \mu_2$ on the $z$-statistic given by

$$z = \frac{\bar{x}_1 - \bar{x}_2 - (\mu_1 - \mu_2)}{\sqrt{\frac{\sigma_1^2}{n_1} + \frac{\sigma_2^2}{n_2}}}.$$

Under these assumptions $z$ has a $N(0, 1)$ distribution. Therefore a $1 - \alpha$ confidence interval for $\mu_1 - \mu_2$ is given by

$$\bar{x}_1 - \bar{x}_2 \pm \sqrt{\frac{\sigma_1^2}{n_1} + \frac{\sigma_2^2}{n_2}} z^*$$

where $z^*$ is the $1 - \alpha/2$ percentile of the $N(0, 1)$ distribution. Further we can test $H_0 : \mu = \mu_0$ against the alternative $H_a : \mu \neq \mu_0$ by computing the $P$-value $P(|Z| > |z_0|) = 2P(Z > z_0)$ where $Z$ is distributed $N(0, 1)$ and $z_0$ is the observed value of the $z$-statistic. These inferences are also appropriate without normality provided $n_1$ and $n_2$ are large and we have reasonable values for $\sigma_1$ and $\sigma_2$. These inferences are easily carried out using Minitab commands we have already discussed, in particular **cdf** and **invcdf** with the **normal** subcommand.

In general, however, we will not have available suitable values of $\sigma_1$ and $\sigma_2$ or large samples and will have to use the two-sample analogs of the single-sample $t$ procedures just discussed. This is acceptable provided of course that we have checked for and agreed that it is reasonable to assume that both samples are from normal distributions. These procedures are based on the two-sample $t$-statistic given by

$$t = \frac{\bar{x}_1 - \bar{x}_2 - (\mu_1 - \mu_2)}{\sqrt{\frac{s_1^2}{n_1} + \frac{s_2^2}{n_2}}}$$

where we have replaced the population standard deviations by their sample estimates . The exact distribution of this statistic does not have a convenient form but of course we can always simulate its distribution. Actually it is typical to use an approximation to the distribution of this statistic based on a Student distribution. See the discussion in IPS on this and use **help** to get more details.

The **twosample** and **twot** commands are available for computing inference procedures based on $t$. Each of these commands computes confidence intervals for the difference of the means and computes $P$-values for tests of significance concerning the difference of means. The only difference between these commands is that with **twosample** the two samples are in individual columns while with **twot** the samples are in a single column with subscripts indicating group membership in a second column.

For example, suppose that we have the data for Example 7.14 in a worksheet with the control sample in C1 and the treatment sample in C2. Then the **twosample** command in

```
MTB > twosample .95 c1 c2
Two sample T for C1 vs C2
  N  Mean  StDev   SE  Mean
C1   23   41.5  17.2   3.6
C2   21   51.5  11.0   2.4
95% CI for mu C1 - mu C2:  ( -18.7, -1.2)
T-Test mu C1 = mu C2 (vs not =):  T = -2.32 P = 0.026 DF = 37
```

gives a 95% confidence for the difference in the means $\mu_1 - \mu_2$ as $(-18.7, -1.2)$ and calculates the $P$-value .026 for the test of $H_0 : \mu_1 - \mu_2 = 0$ versus the alternative $H_a : \mu_1 - \mu_2 \neq 0$.

Alternatively we can place both samples in a single column C1, with the control observations followed by the treatment observations and with subscripts in C2 indicating which sample each observation came from. Then the **twot** command in

```
MTB > set c2
DATA> 23(1)
DATA> 21(2)
DATA> end
MTB > twot 95 c1 c2
Two sample T for C1
C2    N   Mean   StDev  SE Mean
1    23   41.5   17.2     3.6
2    21   51.5   11.0     2.4
95% CI for mu (1) - mu (2):  ( -18.7, -1.2)
```

```
   T-Test mu(1) = mu(2) (vs not =):   T = -2.32 P = 0.026 DF = 37
```
gives exactly the same output.

The general syntax of the **twosample** command is

**twosample** V $E_1E_2$

where V is the confidence level and is any value between 1 and 99.99 and $E_1$, $E_2$ are columns of data containing the two samples. The general syntax of the **twot** command is

**twot** V $E_1E_2$

where V is the confidence level and is any value between 1 and 99.99 and $E_1$, $E_2$ are columns of data with $E_1$ containing the samples and $E_2$ containing the subscripts.

The **alternative** subcommand is available with both **twosample** and **twot** if we wish to conduct one-sided tests. Also the subcommand **pooled** is available if we feel we can assume that $\sigma_1 = \sigma_2 = \sigma$ and want to pool both samples together to estimate the common $\sigma$. Pooling is usually unnecessary and is not recommended.

Simulation can be used to approximate the power of the two sample $t$ test. Note that in this case we must specify the difference $\mu_1 - \mu_2$ as well as $\sigma_1 = \sigma_2$. See Exercise II.7.8 for further details on this.

In the window environment two-sample $t$-tests can be carried out using Stat ▶ Basic Statistics ▶ 2-Sample t and filling in the dialog box appropriately. Also in the window environment exact power calculations can be carried under the assumption of sampling from a normal distribution using Power and Sample Size ▶ 2-Sample t and filling in the dialog box appropriately. Further the minimum sample size required to guarantee a given power at a prescribed difference $|\mu_1 - \mu_2|$, and assuming a common standard deviation $\sigma$, can be obtained using this command.

## 7.6   The $F$-Distribution

If $X_1$ is distributed $Chisquare(k_1)$ independent of $X_2$ distributed $Chisquare(k_2)$, then

$$T = \frac{X_1/k_1}{X_2/k_2}$$

is distributed according to the $F(k_1, k_2)$-distribution. The value $k_1$ is called the *numerator degrees of freedom* and the value $k_2$ is called the *denominator degrees of freedom*. There are Minitab commands that assist in carrying out computations for this distribution.

The values of the density curve for the $F(k_1, k_2)$ distribution can be obtained using the **pdf** command with the **F** subcommand. For example, the command

```
MTB > pdf c1 c2;
SUBC> F 4 6.
```

calculates the value of the $F(4, 6)$ density curve at each value in C1 and stores these values in C2. This is useful for plotting the density curve. The **cdf**, **invcdf** and **random** commands can also be used with the **F** subcommand in this way to obtain values of the $F(k_1, k_2)$ cumulative distribution function and inverse distribution function and to obtain random samples from the distribution respectively.

There are a number of applications of the $F$-distribution some of which arise later in the book but we mention one here. In particular, if $x_{11}, \ldots x_{1n_1}$ is a sample from the $N(\mu_1, \sigma_1)$ distribution and $x_{12}, \ldots x_{1n_2}$ a sample from the $N(\mu_2, \sigma_2)$ distribution, then

$$F = \frac{s_1^2/\sigma_1^2}{s_2^2/\sigma_2^2}$$

is known to follow a $F(n_1 - 1, n_2 - 1)$. As explained in IPS this fact is used as a basis for inference about the ratio $\sigma_1/\sigma_2$, i.e. confidence intervals and tests of significance and in particular testing for equality of variances between the samples. Because of the nonrobustness of these inferences to small deviations from normality these inferences are not recommended.

In the window environment the density curve, cumulative distribution function and inverse distribution function can be evaluated via Çalc ► Probability Distributions ►F while random samples are obtained using Çalc ► Random Data ► F.

## 7.7   Exercises

When the data for an exercise come from an exercise in IPS, the IPS exercise number is given in parentheses ( ). All computations in these exercises are to be carried out using Minitab and the exercises are designed to ensure that you have a reasonable understanding of the Minitab material in this chapter. More generally you should be using Minitab to do all the computations and plotting required for the problems in IPS.

*If your version of Minitab places restrictions such that the value of the simulation sample size N requested in these problems is not feasible, then substitute a more appropriate value. Be aware, however, that the accuracy of your results is dependent on how large N is.*

1. Plot the $Student(k)$ density curve for $k = 1, 2, 10, 30$ and the $N(01)$ density curve on the interval $(-10, 10)$ using an increment of .1 and compare the plots.

2. Make a table of the values of the cumulative distribution function of the $Student(k)$ distribution for $k = 1, 2, 10, 30$ and the $N(01)$ distribution at the points $-10, -5, -3, -1, 0, 1, 3, 5, 10$. Comment on the values.

3. Make a table of the values of the inverse cumulative distribution function of the $Student(k)$ distribution for $k = 1, 2, 10, 30$ and the $N(01)$ distribution at the points .0001, .001, .01, .1, .25, .5. Comment on the values.

4. Simulate $N = 1000$ values from $Z$ distributed $N(0, 1)$ and $X$ distributed $Chisquare(3)$ and plot a histogram of $T = Z/\sqrt{X/3}$ using the cutpoints $-10, -9, ..., 9, 10$. Generate a sample of $N = 1000$ values directly from the $Student(3)$ distribution, plot a histogram with the same cutpoints and compare the two histograms.

5. Carry out a simulation with $N = 1000$ to verify that the 95% confidence interval based on the $t$-statistic covers the true value of the mean 95% of the time when taking samples of size 5 from the $N(4, 2)$ distribution.

6. Generate a sample of 50 from the $N(10, 2)$ distribution. Compare the 95% confidence intervals obtained via the **tinterval** and **zinterval** commands using the sample standard deviation as an estimate of $\sigma$.

7. Carry out a simulation with $N = 1000$ to approximate the power of the $t$-test at $\mu_1 = 1, \sigma_1 = 2$ for testing $H_0 : \mu = 0$ versus the alternative $H_a : \mu \neq 0$ at level $\alpha = .05$ based on a sample of 5 from the normal distribution.

8. Carry out a simulation with $N = 1000$ to approximate the power of the two sample $t$-test at $\mu_1 = 1, \sigma_1 = 2, \mu_2 = 2, \sigma_1 = 3$ for testing $H_0 : \mu_1 - \mu_2 = 0$ versus the alternative $H_a : \mu_1 - \mu_2 \neq 0$ at level $\alpha = .05$ based on a sample of 5 from the $N(\mu_1, \sigma_1)$ distribution and a sample of size 8 from the $N(\mu_2, \sigma_2)$ distribution. Use the conservative rule when choosing the degrees of freedom for the approximate test, i.e. the smaller of $n_1 - 1$ and $n_2 - 1$.

9. If $Z$ is distributed $N(\mu, 1)$ and $X$ is distributed $Chisquare(k)$ independent of $Z$, then

$$Y = \frac{Z}{\sqrt{X/k}}$$

is distributed according to a *noncentral Student(k)* distribution with noncentrality $\mu$. Simulate samples of $N = 1000$ from this distribution with $k = 5$ and $\mu = 0, 1, 5, 10$. Plot the samples in histograms with cutpoints $-20, -19, ..., 19, 20$ and compare these plots.

10. Redo the power simulation in Exercise II.7.7 using an appropriate noncentral Student distribution.

# Chapter 8

# Inference for Proportions

This chapter is concerned with inference methods for a proportion $p$ and for the comparison of two proportions $p_1$ and $p_2$. Proportions arise from measuring a binary-valued categorical variable on population elements such as gender in human populations. For example, $p$ might be the proportion of females in a given population or we might want to compare the proportion $p_1$ of females in population 1 with the proportion $p_2$ of females in population 2. The need for inference arises as we base our conclusions about the values of these proportions on samples from the populations rather than measuring every element in the population. For convenience we will denote the values assumed by the binary categorical variables as 1 and 0 where 1 indicates the presence of a characteristic and 0 indicates its absence.

## 8.1  Inference for a Single Proportion

Suppose that $x_1, \dots, x_n$ is a sample from a population where the variable is measuring the presence or absence of some trait by a 1 or 0 respectively. Let $\hat{p}$ be the proportion of 1's in the sample. This is the estimate of the true proportion $p$. For example, the sample could arise from coin tossing where 1 denotes head and 0 tail and $\hat{p}$ is the proportion of heads while $p$ is the probability of head. If the population we are sampling from is finite, then strictly speaking the sample elements are not independent. But if the population size is large relative to the sample size $n$, then independence is a reasonable approximation and this is necessary for the methods of this chapter. So we will consider $x_1, \ldots, x_n$ as a sample from the $Bernoulli(p)$ distribution.

The standard error of the estimate $\hat{p}$ is $\sqrt{\hat{p}(1-\hat{p})/n}$ and since $\hat{p}$ is an average the central limit theorem gives that

$$z = \frac{\hat{p} - p}{\sqrt{\frac{\hat{p}(1-\hat{p})}{n}}}$$

is approximately $N(0,1)$ for large $n$. This leads to the approximate $1 - \alpha$ confidence interval given by $\hat{p} \pm \sqrt{\hat{p}(1-\hat{p})/n}z^*$ where $z^*$ is the $1 - \alpha/2$ percentile of the $N(01)$ distribution. This interval can be easily computed using Minitab commands. For example, in Example 8.2 in IPS the probability of heads was estimated by Count Buffon as $\hat{p} = .194$ on the basis of a sample of $n = 4040$ tosses. The commands

```
MTB > let k1=.5069
MTB > let k2=sqrt(k1*(1-k1)/4040)
MTB > invcdf .95;
SUBC> normal 0 1.
Normal with mean = 0 and standard deviation = 1.00000
  P( X <= x)          x
      0.9500    1.6449
MTB > let k3=k1-1.6449*k2
MTB > let k4=k1+1.6449*k2
MTB > print k3 k4
K3 0.493962
K4 0.519838
```

compute an approximate 90% confidence interval for $p$ to be $(0.493962, 0.519838)$.

To test a null hypothesis $H_0 : p = p_0$ we make use of the fact that under the null hypothesis the statistic

$$z = \frac{\hat{p} - p_0}{\sqrt{\frac{p_0(1-p_0)}{n}}}$$

is approximately $N(0,1)$. To test $H_0 : p = p_0$ versus $H_a : p \neq p_0$ we compute $P(|Z| > |z|) = 2P(Z > |z|)$ where $Z$ is distributed $N(0,1)$. For example, in

Example 8.2 of IPS suppose we want to test $H_0 : p = .5$ versus $H_a : p \neq .5$. Then the commands

```
MTB > let k1=.5069
MTB > let k2=sqrt(.5*.5/4040)
MTB > let k3=absolute((k1-.5)/k2)
MTB > cdf k3;
SUBC> normal 0 1.
Normal with mean = 0 and standard deviation = 1.00000
      x        P( X <= x)
  0.8771          0.8098
MTB > let k4=1-.8098
MTB > let k5=2*k4
MTB > print k5
K5 0.380400
```

compute the $P$-value of this two-sided test to be .3804. The formulas provided in IPS for computing the $P$-values associated with one-sided tests are also easily implemented in Minitab.

We can simulate to estimate the power of these tests. For example, suppose we want to compute the power of the test for $H_0 : p = .5$ versus $H_a : p \neq .5$ at level $\alpha = .05$ at $n = 10, p = .4$. Then the commands

```
MTB > random 1000 c1;
SUBC> binomial 10 .4.
MTB > let c2=c1/10
MTB > let c3=absolute((c2-.5)/sqrt(.25/10))
MTB > invcdf .975;
SUBC> normal 0 1.
Normal with mean = 0 and standard deviation = 1.00000
  P( X <= x)         x
    0.9750    1.9600
MTB > let c4=c3>1.96
MTB > let k1=mean(c4)
MTB > let k2=sqrt(k1*(1-k1)/n(c4))
MTB > print k1 k2
K1 0.0450000
K2 0.00655553
```

approximate this power as .045 with standard error .007. So the test is not very powerful and, as indicated by the discussion in IPS, should not be used. By contrast at $n = 100, p = .4$ the power is approximated to be .543 with standard error .016. For large values of $n$ the normal approximation must be use to calculate the power.

In the window environment the confidence interval and test of significance can be carried out using Stat ▶ Basic Statistics ▶ 1 Proportion and filling in the dialog box. Power calculations and minimum sample sizes to achieve a prescribed power can be obtained using Power and Sample Size ▶ 1 Proportion.

## 8.2   Inference for Two Proportions

Suppose that $x_{11}, \ldots, x_{n_1 1}$ is a sample from population 1 and $x_{12}, \ldots, x_{n_2 2}$ is a sample from population 2 where the variable is measuring the presence or absence of some trait by a 1 or 0 respectively. We assume then that we have a sample of $n_1$ from the *Bernoulli*$(p_1)$ distribution and a sample of $n_2$ from the *Bernoulli*$(p_2)$ distribution. Suppose that we want to make inferences about the difference in the proportions $p_1 - p_2$. Let $\hat{p}_i$ be the proportion of 1's in the *ith* sample.

The central limit theorem gives that

$$z = \frac{\hat{p}_1 - \hat{p}_2 - (p_1 - p_2)}{\sqrt{\frac{\hat{p}_1(1-\hat{p}_1)}{n_1} + \frac{\hat{p}_2(1-\hat{p}_2)}{n_2}}}$$

is approximately $N(0,1)$ for large $n_1$ and $n_2$. This leads to the approximate $1 - \alpha$ confidence interval given by

$$\hat{p}_1 - \hat{p}_2 \pm \sqrt{\frac{\hat{p}_1(1-\hat{p}_1)}{n_1} + \frac{\hat{p}_2(1-\hat{p}_2)}{n_2}} z^*$$

where $z^*$ is the $1 - \alpha/2$ percentile of the $N(01)$ distribution. This interval can be easily computed using Minitab commands just as we did for a confidence interval for a single proportion in II.8.1.

To test a null hypothesis $H_0 : p_1 = p_2$ we use the fact that under the null hypothesis the statistic

$$z = \frac{\hat{p}_1 - \hat{p}_2}{\sqrt{\hat{p}(1-\hat{p})\left(\frac{1}{n_1} + \frac{1}{n_2}\right)}}$$

is approximately $N(0,1)$ for large $n_1$ and $n_2$ where $\hat{p} = (n_1\hat{p}_1 + n_2\hat{p}_2)/(n_1 + n_2)$ is the estimate of the common value of the proportion when the null hypothesis is true. To test $H_0 : p_1 = p_2$ versus $H_a : p_1 \neq p_2$ we compute $P(|Z| > |z|) = 2P(Z > |z|)$ where $Z$ is distributed $N(0,1)$. For example, in Example 8.9 of IPS suppose that we want to test the $H_0 : p_1 = p_2$ versus $H_a : p_1 \neq p_2$ where $n_1 = 7180, \hat{p}_1 = .227, n_2 = 9916, \hat{p}_2 = .170$. Then the commands

```
MTB > let k1=(7180*.227+9916*.170)/(7180+9916)
MTB > let k2=sqrt(k1*(1-k1)*(1/7180+1/9916))
MTB > let k3=absolute((.227-.170)/k2)
MTB > cdf k3;
SUBC> normal 0 1.
Normal with mean = 0 and standard deviation = 1.00000
       x       P( X <= x)
   9.3034         1.0000
```

compute the $P$-value as $2(1-1) = 0$ so we would definitely reject.

Approximate power calculations can be carried out by simulating $N$ pairs of values from the $Binomial(n_1, p_1)$ and $Binomial(n_2, p_2)$ distribution. For example, the commands

```
MTB > random 1000 c1;
SUBC> binomial 40 .3.
MTB > random 1000 c2;
SUBC> binomial 50 .5.
MTB > let c3=c1/40
MTB > let c4=c2/50
MTB > let c5=(40*c3+50*c4)/(40+50)
MTB > let c6=absolute((c3-c4)/sqrt(c5*(1-c5)*(1/40+1/50)))
MTB > invcdf .975;
SUBC> normal 0 1.
Normal with mean = 0 and standard deviation = 1.00000
   P( X <= x)          x
```

```
      0.9750      1.9600
MTB > let c7=c6>1.96
MTB > let k1=mean(c7)
MTB > let k2=sqrt(k1*(1-k1)/n(c7))
MTB > print k1 k2
K1 0.472000
K2 0.0157866
```

approximate the power of $H_0 : p_1 = p_2$ versus $H_a : p_1 \neq p_2$ at level $\alpha = .05$ at $n_1 = 40, p_1 = .3, n_2 = 50, p_2 = .5$ to be .472 with standard error .016.

In the window environment the confidence interval and test of significance can be carried out using Stat ▶ Basic Statistics ▶ 2 Proportions and filling in the dialog box. Power calculations and minimum sample sizes to achieve a prescribed power can be obtained using Power and Sample Size ▶ 2 Proportions.

## 8.3   Exercises

When the data for an exercise come from an exercise in IPS, the IPS exercise number is given in parentheses ( ). All computations in these exercises are to be carried out using Minitab and the exercises are designed to ensure that you have a reasonable understanding of the Minitab material in this chapter. More generally you should be using Minitab to do all the computations and plotting required for the problems in IPS.

Don't forget to quote standard errors for any approximate probabilities you quote in the following problems.

1. Carry out a simulation with the $Binomial(40, .3)$ distribution to assess the coverage of the 95% confidence interval for a single proportion.

2. The accuracy of a confidence interval procedure can be assessed by computing *probabilities of covering false values*. Approximate the probabilities of covering the values .1, .2, ...,.9 for the 95% confidence interval for a single proportion when sampling from the $Binomial(20, .5)$ distribution.

3. Approximate the power of the two-sided test for testing $H_0 : p = .5$ at level $\alpha = .05$ at the points $n = 100, p = .1, ..., 9$ and plot the power curve.

4. Carry out a simulation with the $Binomial(40, .3)$ and the $Binomial(50, .4)$ distribution to assess the coverage of the 95% confidence interval for a difference of proportions.

5. Approximate the power of the two-sided test for testing $H_0 : p_1 = p_2$ versus $H_a : p_1 \neq p_2$ at level $\alpha = .05$ at $n_1 = 40, p_1 = .3, n_2 = 50, p_2 = .1, ..., 9$ and plot the power curve.

6. (8.15) It is possible to use the **zinterval** command to construct confidence intervals for a proportion $p$. For this we must put the appropriate number of 1's and 0's in a column and specify $\sigma$ appropriately. Do this problem in IPS using **zinterval**.

7. (8.16) It is possible to use the **ztest** command to carry out tests of significance for a proportion $p$. For this we must put the appropriate number of 1's and 0's in a column and specify $\sigma$ appropriately. Do this problem in IPS using **ztest**.

# Chapter 9

# Inference for Two-way Tables

<u>New Minitab commands discussed in this chapter</u>
**chisquare**

In this chapter inference methods are discussed for comparing the distributions of a categorical variable for a number of populations and for looking for relationships amongst a number of categorical variables defined on a single population. The *chi-square test* is the basic inferential tool and this is implemented in Minitab via the **table** command if the data is in the form of raw incidence data or the **chisquare** command if the data comes in the form of counts.

## 9.1 The TABLE Command

You should recall or reread the discussion of the **table** command in II.2.2 as we will mention here only the additional features related to carrying out the chi-square test. So for example, suppose for 60 cases we have a categorical variable in C1 taking the values 0 and 1 and a categorical variable in C2 taking the values 0, 1 and 2. Then the command

```
MTB > table c1 c2
```

    Rows:  C1     Columns:  C2

|      | 0  | 1  | 2  | All |
|------|----|----|----|-----|
| 0    | 10 | 13 | 11 | 34  |
| 1    | 9  | 10 | 7  | 26  |
| All  | 19 | 23 | 18 | 60  |

```
Cell Contents --
          Count
```

records the counts in the 6 cells of a table with C1 indicating row and C2 indicating column. The variable C1 could be indicating a population with C2 a categorical variable defined on each population (or conversely) or both variables could be defined on a single population.

There is no relationship between the variables – i.e. the variables are *independent* – if and only if the conditional distributions of C2 given C1 are all the same. Alternatively we can express this in terms of the conditional distributions of C1 given C2. In any case we can assess whether or not there is a relationship by comparing the conditional distributions of the columns given the rows. Of course there will be differences in these conditional distributions simply due to sampling error. Whether or not these differences are significant is assessed by conducting a chi-square test and this can be carried out using the **chisquare** subcommand to **table**. The command

```
MTB > table c1 c2;
SUBC> rowpercents;
SUBC> chisquare.
```

```
Rows:  C1    Columns:  C2
```

|      | 0     | 1     | 2     | All    |
|------|-------|-------|-------|--------|
| 0    | 29.41 | 38.24 | 32.35 | 100.00 |
|      | 10    | 13    | 11    | 34     |
| 1    | 34.62 | 38.46 | 26.92 | 100.00 |
|      | 9     | 10    | 7     | 26     |
| All  | 31.67 | 38.33 | 30.00 | 100.00 |

```
              19          23          18          60

   Chi-Square = 0.271, DF = 2, P-Value = 0.873

     Cell Contents --
                    % of Row
                    Count
```

computes the conditional distributions given the row variable and then carries out the chi-square test. In this case the *P*-value is .873 which indicates that there is no evidence of a difference among the conditional distributions.

It is possible to cross-tabulate more than two variables and to test simultaneously for mutual statistical independence among the variables using the **chisquare** subcommand. The general syntax of this command is

**table** $E_1$ ... $E_m$;
**chisquare** V.

where $E_1$, ..., $E_m$ are columns containing categorical variables and V is either omitted or takes the value 1, 2 or 3. The value 1 is the default and causes the count to be printed in each cell and can be omitted. The value 2 causes the count and the expected count, under the hypothesis of independence, to be printed in each cell. The value 3 causes the count, the expected count and the standardized residual to be printed in each cell. Note that the square of the standardized residual is that cell's contribution to the chi-square statistic, namely

$$\frac{(\text{observed count in cell } - \text{ expected count in cell })^2}{\text{expected count in cell}}$$

in

$$X^2 = \sum_{\text{cell}} \frac{(\text{observed count in cell } - \text{ expected count in cell })^2}{\text{expected count in cell}}$$

The *P*-value of the chi-square test is obtained by computing the probability

$$P(Y > X^2)$$

where $Y$ follows a *Chisquare* $(k)$ distribution based on an appropriate degrees of freedom $k$ as determined by the table and the model being fitted. When the table has $r$ rows and $c$ columns, and we are testing for independence, then $k = (r-1)(c-1)$. This is an approximate distribution result. Recall that the *Chisquare* $(k)$ distribution was discussed in II.6.5.

In the window environment this analysis can be carried using $\underline{S}$tat ▶ $\underline{T}$ables ▶ $\underline{C}$ross Tabulation and filling in the dialog box appropriately. Recall from II.2.2 that it is also a good idea to plot the conditional distributions as well and this can be carried out using $\underline{G}$raph ▶ $\underline{C}$harts and filling in the dialog box as described there.

## 9.2  The CHISQUARE Command

If you have a cross-tabulation for which the cell counts are already tabulated, then you can use the **chisquare** command on this data to carry out the chisquare test. For example suppose we put the table in Example 9.8 of IPS in columns C1-C3 as

```
MTB > print c1-c3
 Row    C1     C2     C3
  1     51     22     43
  2     92     21     28
  3     68      9     22
```

and then we use the command

```
MTB > chisquare c1-c3
Expected counts are printed below observed counts
          C1      C2      C3      Total
   1      51      22      43       116
        68.75   16.94   30.30
   2      92      21      28       141
        83.57   20.60   36.83
   3      68       9      22        99
        58.68   14.46   25.86
Total    211      52      93       356
Chi-Sq = 4.584 + 1.509 + 5.320 +
  0.850 + 0.008 + 2.119 +
  1.481 + 2.062 + 0.577 = 18.510
DF = 4, P-Value = 0.001
```

on these columns to compute the chi-square test statistic for independence. The statistic takes the value 18.51 in this case and the *P*-value is .001.

The general syntax of the **chisquare** command is

> **chisquare** $E_1$ ... $E_m$

and this computes the expected cell counts, the chi-square statistic and the associated *P*-value for the table in columns $E_1$, ..., $E_m$. Note that there is a limitation on the number of columns; namely we must have $m \leq 7$.

In the window environment this analysis can be carried using Stat ▶ Tables ▶ Chi-Square Test and filling in the dialog box appropriately.

# 9.3  Exercises

When the data for an exercise come from an exercise in IPS, the IPS exercise number is given in parentheses ( ). All computations in these exercises are to be carried out using Minitab and the exercises are designed to ensure that you have a reasonable understanding of the Minitab material in this chapter. More generally you should be using Minitab to do all the computations and plotting required for the problems in IPS.

1. Use Minitab to directly compute the expected frequencies, standardized residuals, chi-square statistic and *P*-value for the hypothesis of independence in the table of Example 9.8 in IPS. Verify your results using the **chisquare** command.

2. (9.17) If it is available in the version of Minitab you are using, plot bar charts of the conditional distributions. Make sure you use the same scale on each plot so that they are comparable.

3. Suppose we have a discrete distribution on the integers $1, \ldots, k$ with probabilities $p_1, \ldots, p_k$. Further suppose we take a sample of $n$ from this distribution and record the counts $f_1, \ldots, f_k$ where $f_i$ records the number of times we observed $i$. Then it can be shown that

$$P(f_1 = n_1, \ldots, f_k = n_k) = \frac{n!}{n_1! \cdots n_k!} p_1^{n_1} \cdots p_k^{n_k}$$

when the $n_i$ are nonnegative integers that sum to $n$. This is called the *Multinomial*$(n, p_1, \ldots, p_k)$ distribution and it is a generalization of the

*Binomial*$(n, p)$ distribution. It is the relevant distribution for describing the counts in cross-tabulations. For $k = 4, p_1 = p_2 = p_3 = p_4 = .25, n = 3$ calculate these probabilities and verify that it is a probability distribution. Recall that the **gamma** command (see Appendix B.1) can be used to evaluate factorials such as $n!$ and also $0! = 1$.

4. Calculate $P(f_1 = 3, f_2 = 5, f_3 = 2)$ for the *Multinomial*$(10, .2, .5, .3)$ distribution.

5. Generate $(f_1, f_2, f_3)$ from the *Multinomial*$(1000, .2, .4, .4)$ distribution. Hint: Generate a sample of 1000 from the discrete distribution on 1, 2, 3 with probabilities .2, .4 , .4 respectively.

# Chapter 10

# Inference for Regression

This chapter deals with inference for the simple linear model. The **regress** command for the fitting of this model was discussed in II.2.1.4 and this material should be recalled or reread at this point. Here we give a list of all the subcommands available with regress and present an example.

## 10.1  Subcommands for REGRESS

The general syntax of the **regress** command for fitting a line is

   **regress** $E_1$ 1 $E_2$

where $E_1$ contains the values of the response variable $y$ and $E_2$ contains the values of the explanatory variable $x$. This provides a fit of the model $y = \beta_0 + \beta_1 x + \epsilon$. The least-squares estimates of $\beta_0$ and $\beta_1$ are denoted $b_0$ and $b_1$ respectively and these are based on the observed data $(x_1, y_1), \ldots, (x_n, y_n)$. We also estimate the standard deviation $\sigma$ by $s$ which equals the square root of the MSE (mean-squared error) for the regression model.

In II.10.2 we will explain in detail the output from the **regress** command. We note, however, that there are a number of subcommands that can be used with **regress** and these are listed and explained. Some of these subcommands are associated with regression concepts that are not treated in

IPS but are appropriate for more advanced treatments. We have indicated these subcommands with (*) and these can be skipped if so desired.

**coefficients** $E_1$ - stores the estimates of the coefficients in column $E_1$.

**constant (noconstant)** - ensures that $\beta_0$ is included in the regression equation while **noconstant** fits the equation without $\beta_0$.

**cookd** $E_1$ - computes *Cook's distance* and stores it in $E_1$. (*)

**dw** - causes the *Durbin-Watson test statistic* to be printed. (*)

**dfits** $E_1$ - stores *dffits in* $E_1$.

**fits** $E_1$ - stores the *fitted values* $\hat{y}$ in $E_1$.

**ghistogram** - causes a histogram of the residuals specified in **rtype** to be plotted.

**gfits** - causes a plot of the residuals specified in **rtype** versus the fitted values to be plotted.

**gnormal** - causes a normal quantile plot of the residuals specified in **rtype** to be plotted.

**gorder** - causes a plot of the residuals specified in **rtype** versus order to be plotted.

**gvariable** $E_1$ - causes a plot of the residuals specified in **rtype** versus the explanatory variable in column $E_1$ to be plotted.

**hi** $E_1$ - computes the *leverages* and stores these in $E_1$. (*)

**mse** $E_1$ - stores the mean squared error in constant $E_1$.

**predict** $E_1 \ldots E_k$ - ($k$ is the number of explanatory variables where $k = 1$ with simple linear regression) computes and prints the predicted values at $E_1$, ..., $E_k$ where these are columns of the same length or constants with $E_i$ corresponding to the *ith* explanatory variable. Also prints the estimated standard deviations of these values, confidence intervals for these values and prediction intervals. The subcommand **predict** in turn has a number of subcommands.

>   **confidence** V - V specifies the level for the confidence intervals.
>
>   **pfits** $E_1$ - stores the predicted values in $E_1$.
>
>   **psdfits** $E_1$ - stores the estimated standard deviations of the predicted values in $E_1$.
>
>   **climits** $E_1$ $E_2$ - stores the lower and upper confidence limits for the predicted values in $E_1$ and $E_2$ respectively.
>
>   **plimits** $E_1$ $E_2$ - stores the lower and upper prediction limits for the predicted values in $E_1$ and $E_2$ respectively.

**pure** - carries out a *pure lack of fit test* (when there are replicates of observations). (*)

**residuals** $E_1$ - stores the regular residuals in $E_1$.

**rmatrix** $E_1$ - stores the $R$ matrix in the matrix $E_1$ where $X = QR$ is the *Cholesky decomposition* of the design matrix (see Appendix D). (*)

**rtype** V - indicates what type of residuals are to be used in the plotting subcommands where V = 1 is the default and specifies regular residuals, V = 2 specifies standardized residuals and V = 3 specifies Studentized deleted residuals.

**sresiduals** $E_1$ - stores the standardized residuals – the residuals divided by their estimated standard deviations – in $E_1$.

**tolerance** $V_1$ $V_2$ - this is applicable in multiple regression contexts as discussed in Chapter 11. Minitab automatically removes explanatory variables from a model when they are highly correlated with other explanatory variables or if they are nearly constant. The value $V_1$ controls the degree of correlation permitted before explanatory variables are removed and the value $V_2$ controls the degree of constancy allowed. These can be lowered from their default values if variables you want kept are being removed. The default values are $10^{-18}$ and $2 \times 10^{-21}$ respectively. See **help** for more details. (*)

**tresiduals** $E_1$ - stores the Studentized residuals in $E_1$.

**vif** - prints the *variance inflation factor* associated with each explanatory variable. (*)

**weights** $E_1$ - the weights in column $E_1$ are used to carry out a weighted regression; i.e. the vector of least-squares estimates is given by $(X^tWX)^{-1} X^tWy$ where $W$ is a diagonal matrix with the weights along the diagonal (see Appendix D). (*)

**xpinv** $E_1$ - stores the matrix $(X^tX)^{-1}$ in the matrix $E_1$ where $X$ is the design matrix (see Appendix D). If the **weight** subcommand is used then if the weights are along the diagonal of the matrix $W$ the matrix $(X^tWX)^{-1}$ is stored in the matrix $E_1$.(*)

In the window environment regression analysis can be carried out using Stat ▶ Regression ▶ Regression and filling in the dialog box appropriately. The regression as well as a plot with the least-squares line overlaid can be obtained via Stat ▶ Regression ▶ Fitted Line Plot and residual plots can be obtained using Stat ▶ Regression ▶ Residual Plots provided you have saved the residuals.

## 10.2  Example

We illustrate the use of the **regress** command and the subcommands using Example 10.8 in IPS. For this we have 4 data points

$$
\begin{aligned}
(x_1, y_1) &= (1966, 73.1) \\
(x_2, y_2) &= (1976, 88.0) \\
(x_3, y_3) &= (1986, 119.4) \\
(x_4, y_4) &= (1996, 127.1)
\end{aligned}
$$

where $x$ is year and $y$ is yield in bushels per acre. Suppose that we gave $x$ the name year and $y$ the name yield. Then the commands

```
MTB > regress 'yield' 1 'year';
SUBC> coefficients c3;
SUBC> mse k1;
SUBC> fits c4;
SUBC> sresiduals c5;
SUBC> rtype 2;
SUBC> gnormal;
SUBC> gvariable 'year';
SUBC> predict 2006;
SUBC> climits c6 c7;
SUBC> plimits c8 c9.
```

```
Regression Analysis
The regression equation is
yield = - 3729 + 1.93 year
```

| Predictor | Coef | StDev | T | P |
|-----------|--------|-------|-------|-------|
| Constant | -3729.4 | 606.6 | -6.15 | 0.025 |
| year | 1.9340 | 0.3062 | 6.32 | 0.024 |

```
S = 6.847 R-Sq = 95.2% R-Sq(adj) = 92.8%
```

```
Analysis of Variance
```

| Source | DF | SS | MS | F | P |
|--------|-----|--------|--------|-------|-------|
| Regression | 1 | 1870.2 | 1870.2 | 39.89 | 0.024 |
| Residual Error | 2 | 93.8 | 46.9 | | |
| Total | 3 | 1963.9 | | | |

```
Predicted Values
```

```
   Fit   StDev Fit       95.0% CI          95.0% PI
  150.25        8.39  ( 114.17, 186.33)  ( 103.67, 196.83) X
 X denotes a row with X values away from the center
```

give the least-squares line as $y = -3729 + 1.93x$. Further the standard error of $b_0 = -3729.4$ is 606.6, the standard error of $b_1 = 1.934$ is 0.3062, the $t$-statistic for testing $H_0 : \beta_0 = 0$ versus $H_a : \beta_0 \neq 0$ is -6.15 with $P$-value 0.025 and the $t$-statistic for testing $H_0 : \beta_1 = 0$ versus $H_a : \beta_1 \neq 0$ is 6.32 with $P$-value 0.024. The estimate of $\sigma$ is $s = 6.847$ and the squared correlation is $R^2 = .952$ indicating that 95% of the observed variation in $y$ is explained by the changes in $x$. The Analysis of Variance table indicates that the $F$-statistic for testing $H_0 : \beta_1 = 0$ versus $H_a : \beta_1 \neq 0$ is 39.89 with $P$-value 0.024 and the MSE is 46.9. The predicted value at $x = 2006$ is 150.25 with standard error 8.39 and the 95% confidence and prediction intervals for this quantity are (114.17, 186.33) and (103.67, 196.83) respectively. A high resolution normal quantile plot of the standardized residuals and a high resolution plot of the standardized residuals against the explanatory variable are also plotted but not reproduced here. Note the use of **rtype** 2 to select the standardized residuals for these plots. Depending on the version of Minitab you are using these plots may not be available to you via these subcommands but of course can be plotted directly as we described in II.2.1.4. Further we have stored the values of $b_0$ and $b_1$ in column C3, the MSE in constant K1, the fitted values in C4, the standardized residuals in C5, the lower and upper end-points of the 95% confidence intervals for $\hat{y}$ at $x = 2006$ in C6 and C7 and the lower and upper end-points of the 95% prediction intervals for $\hat{y}$ at $x = 2006$ in C8 and C9. All of these quantities are available for further use. For example, suppose we want a 95% confidence interval for $\beta_1$. Then the commands

```
MTB > invcdf .975;
SUBC> student 2.
Student's t distribution with 2 DF
 P( X <= x)         x
    0.9750      4.3027
MTB > let k2=4.3027*.3062
MTB > let k3=c3(2)-k2
MTB > let k4=c3(2)+k2
MTB > print k3 k4
K3 0.616513
K4 3.25149
```

gives this interval as $(0.617, 3.251)$.

## 10.3   Exercises

When the data for an exercise come from an exercise in IPS, the IPS exercise
number is given in parentheses ( ). All computations in these exercises are
to be carried out using Minitab and the exercises are designed to ensure that
you have a reasonable understanding of the Minitab material in this chapter.
More generally you should be using Minitab to do all the computations and
plotting required for the problems in IPS.

1. In C1 place the $x$ values $-3.0$, $-2.5$, $-2.0$, ..., 2.5, 3.0. In C2 store a
   sample of 13 from the error $\epsilon$ where $\epsilon$ is distributed $N(0, 2)$. In C3 store
   the values $y = \beta_0 + \beta_1 x + \epsilon = 1 + 3x + \epsilon$. Calculate the least-squares
   estimates of $\beta_0$ and $\beta_1$ and the estimate of $\sigma^2$. Repeat this example
   but take 5 observations at each value of $x$. Compare the estimates from
   the two situations and their estimated standard deviations.

2. In C1 place the $x$ values $-3.0$, $-2.5$, $-2.0$, ..., 2.5, 3.0. In C2 store
   a sample of 13 from the error $\epsilon$ where $\epsilon$ is distributed $N(0, 2)$. In C3
   store the values $y = \beta_0 + \beta_1 x + \epsilon = 1 + 3x + \epsilon$. Plot the least-squares
   line. Now repeat your computations twice after changing the first $y$
   observation to 20 and then to 50 and make sure the scales on all the
   plots are the same. What effect do you notice?

3. In C1 place the $x$ values $-3.0$, $-2.5$, $-2.0$, ..., 2.5, 3.0. In C2 store
   a sample of 13 from the error $\epsilon$ where $\epsilon$ is distributed $N(0, 2)$. In C3
   store the values $y = \beta_0 + \beta_1 x + \epsilon = 1 + 3x + \epsilon$. Plot the standardized
   residuals in a normal quantile plot against the fitted values and against
   the explanatory variable. Repeat this but in C3 place the values of
   $y = 1 + 3x - 5x^2 + \epsilon$. Compare the residual plots.

4. In C1 place the $x$ values $-3.0$, $-2.5$, $-2.0$, ..., 2.5, 3.0. In C2 store
   a sample of 13 from the error $\epsilon$ where $\epsilon$ is distributed $N(0, 2)$. In C3
   store the values $y = \beta_0 + \beta_1 x + \epsilon = 1 + 3x + \epsilon$. Plot the standardized
   residuals in a normal quantile plot against the fitted values and against
   the explanatory variable. Repeat this but in C2 place the values of a
   sample of 13 from the $Student(1)$ distribution. Compare the residual
   plots.

5. In C1 place the $x$ values $-3.0$, $-2.5$, $-2.0$, ..., 2.5, 3.0. In C2 store a sample of 13 from the error $\epsilon$ where $\epsilon$ is distributed $N(0, 2)$. In C3 store the values $y = \beta_0 + \beta_1 x + \epsilon = 1 + 3x + \epsilon$. Calculate the predicted values and the lengths of .95 confidence and prediction intervals for this quantity at $x = .1, 1.1, 2.1, 3.5, 5, 10$ and 20. Explain the effect that you observe.

6. In C1 place the $x$ values $-3.0$, $-2.5$, $-2.0$, ..., 2.5, 3.0. In C2 store a sample of 13 from the error $\epsilon$ where $\epsilon$ is distributed $N(0, 2)$. In C3 store the values $y = \beta_0 + \beta_1 x + \epsilon = 1 + 3x + \epsilon$. Calculate the least-squares estimates and their estimated standard deviations. Repeat this but for C1 now take the $x$ values to be 12 values of -3 and one value of 3. Compare your results and explain them.

# Chapter 11

# Multiple Regression

**New Minitab command discussed in this chapter**

brief

In this chapter we discuss *multiple regression*; i.e. we have a single numeric response variable $y$ and $k > 1$ explanatory variables $x_1, \ldots, x_k$. There are no real changes in the behavior of the **regress** command and in fact the descriptions we gave in Chapter 10 of the subcommands apply as well to this chapter. We present an example of a multiple regression analysis using Minitab. We introduce the **brief** command which allows for a degree of control over the amount of output from the **regress** command. In Chapter 15 we show how a *logistic regression* – a regression where the response variable $y$ is binary-valued – can be carried out using Minitab.

In the window environment regression analysis can be carried out using Stat ▶ Regression ▶ Regression and filling in the dialog box appropriately. Residual plots can be obtained using Stat ▶ Regression ▶ Residual Plots provided you have saved the residuals. Also available in Minitab are *stepwise regression* using Stat ▶ Regression ▶ Regression ▶ Stepwise and *best subsets regression* using Stat ▶ Regression ▶ Regression ▶ Best Subsets.

## 11.1 Example

We consider a generated multiple regression example to illustrate the use of the **regress** command in this context. Suppose that $k = 2$ and $y = \beta_0 + \beta_1 x_1 + \beta_2 x_2 + \varepsilon = 1 + 2x_1 + 3x_2 + \varepsilon$ where $\epsilon$ is distributed $N(0, \sigma)$ with $\sigma = 1.5$. We generate a sample of 16 from the $N(0, 1.5)$ distribution and place these values in C1. Then in C2 we store the values of $x_1$ and in C3 store the values of $x_2$. Suppose that these variables take every possible combination of $x_1 = -1, -.5, .5, 1$ and $x_2 = -2, -1, 1, 2$. In C4 we place the values of the response variable $y$. We then proceed to analyze this data as if we didn't know the values of $\beta_0, \beta_1, \beta_2$ and $\sigma$. The commands

```
MTB > regress c4 2 c2 c3;
SUBC> coefficients c5;
SUBC> rtype 2;
SUBC> gnormal;
SUBC> gvariable C2 C3;
SUBC> predict 0 0;
SUBC> climits c6 c7;
SUBC> plimits c8 c9.
```

The regression equation is
C4 = 1.00 + 2.38 C2 + 2.50 C3

| Predictor | Coef | StDev | T | P |
|---|---|---|---|---|
| Constant | 1.0014 | 0.3307 | 3.03 | 0.010 |
| C2 | 2.3807 | 0.4183 | 5.69 | 0.000 |
| C3 | 2.4964 | 0.2092 | 11.94 | 0.000 |

S = 1.323 R-Sq = 93.1% R-Sq(adj) = 92.0%

Analysis of Variance

| Source | DF | SS | MS | F | P |
|---|---|---|---|---|---|
| Regression | 2 | 305.95 | 152.98 | 87.42 | 0.000 |
| Residual Error | 13 | 22.75 | 1.75 | | |
| Total | 15 | 328.70 | | | |

| Source | DF | Seq SS |
|---|---|---|
| C2 | 1 | 56.68 |
| C3 | 1 | 249.28 |

Unusual Observations

| Obs | C2 | C4 | Fit | StDev Fit | Residual | St Resid |
|---|---|---|---|---|---|---|

```
  6  -0.50  -5.574  -2.685     0.444     -2.889     -2.32R
  R denotes an observation with a large standardized residual
  Predicted Values
     Fit    StDev Fit        95.0% CI            95.0% PI
   1.001         0.331     ( 0.287, 1.716)    ( -1.944, 3.947)
```

give the least-squares equation as $y = 1.00 + 2.38x_1 + 2.50x_2$. For example, the estimate of $\beta_1$ is $b_1 = 2.3807$ with standard error 0.4183 and the $t$-statistic for testing $H_0 : \beta_1 = 0$ versus $H_a : \beta_1 \neq 0$ is 5.69 with $P$-value 0.000. The estimate of $\sigma$ is $s = 1.323$ and $R^2 = .931$. The Analysis of Variance table indicates that the $F$-statistic for testing $H_0 : \beta_1 = \beta_2 = 0$ versus $H_a : \beta_1 \neq 0$ or $\beta_2 \neq 0$ takes the value 87.42 with $P$-value 0.000 so we would definitely reject the null hypothesis. Also the MSE is given as 1.75.

The table after the Analysis of Variance table is called the *sequential Analysis of Variance table* and is used when we want to test whether or not explanatory variables are in the model in a prescribed order. For example, the table which contains the rows labeled C2 and C3 allows for the testing of the sequence of hypotheses $H_0 : \beta_2 = 0$ versus $H_a : \beta_2 \neq 0$ and if we reject this (and only if we do) then testing the hypothesis $H_0 : \beta_1 = 0$ versus $H_a : \beta_1 \neq 0$. To test these hypotheses we first compute $F = 249.28/s^2 = 249.28/1.75 = 142.45$ and then compute the $P$-value $P(F(1, 13) > 142.45) = 0.00$ and so we reject and go no further. If we had not rejected this null hypothesis then the second null hypothesis would be tested in exactly the same way. Obviously the order in which we put variables into the model matters with these sequential tests. Sometimes it is clear how to do this; e.g. in fitting a quadratic model $y = \beta_0 + \beta_1 x + \beta_2 x^2 + \varepsilon$ we put $x_1 = x$ and $x_2 = x^2$ and test for the existence of the quadratic term first and, if no quadratic term is found, then test for the existence of the linear term. Sometimes the order for testing is not as clear and the sequential tests are not as appropriate.

We stored the values of the least-squares estimates in C5 and these are available for forming confidence intervals. For example, the commands

```
MTB > invcdf .95;
SUBC> student 13.
Student's t distribution with 13 DF
  P( X <= x) x
  0.9500 1.7709
MTB > let k1=1.7709*.2092
```

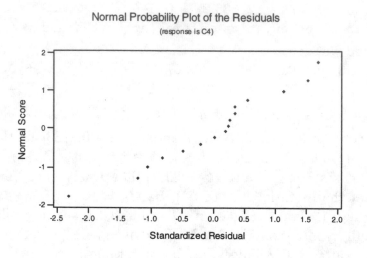

Figure 11.1: Normal quantile plot of standardized residuals.

```
MTB > let k2=c5(3)-k1
MTB > let k3=c5(3)+k1
MTB > print k2 k3
K2  2.12590
K3  2.86685
```

compute a 90% confidence interval for $\beta_2$ as $(2.126, 2.869)$ which we note does not cover the true value in this case.

The program computed the predicted value at $x_1 = 0, x_2 = 0$ as 1.001 with standard error .331 and as well the 95% confidence and prediction intervals given by $(0.287, 1.716)$ and $(-1.944, 3.947)$ respectively. Further, these limits were stored in the columns C6-C9.

We also plotted a normal quantile plot of the standardized residuals, which we show in Figure 11.1, as well as plots of the standardized residuals against each of the explanatory variables, which we don't show. All of these plots look reasonable although we note that the software has identified the sixth observation as having a large standardized residual even though we *know* that the model is correct. Of course 16 is not very many data points so we can expect inference to be somewhat unreliable.

## 11.2   The BRIEF Command

The **brief** command is used to control the amount of output from **regress** and other Minitab commands. The general syntax of the **brief** command is

   **brief** V

where V is a nonnegative integer that controls the amount of output. For any given command the output is dependent on the specific command although V $= 0$ suppresses all output, for all commands, beyond error messages and warnings. The default level of V is 2. When V $=3$ the **regress** command produces the usual output and in addition prints $x, y, \hat{y}$, the standard deviation of $\hat{y}, y - \hat{y}$ and the standardized residual. When V $= 1$ the regress command gives the same output as when V $= 2$ but the sequential analysis of variance table is not printed. Don't forget that after you set the level of **brief** then this may affect the output of all commands you subsequently type and therefore it may need to be reset.

## 11.3   Exercises

When the data for an exercise come from an exercise in IPS, the IPS exercise number is given in parentheses ( ). All computations in these exercises are to be carried out using Minitab and the exercises are designed to ensure that you have a reasonable understanding of the Minitab material in this chapter. More generally you should be using Minitab to do all the computations and plotting required for the problems in IPS.

1. In C1 place the $x_1$ values $-3.0, -2.5, -2.0, ..., 2.5, 3.0$. In C2 store a sample of 13 from the error $\epsilon$ where $\epsilon$ is distributed $N(0, 2)$. In C3 store the values of $x_2 = x^2$. In C4 store the values of $y = \beta_0 + \beta_1 x_1 + \beta_2 x_2 + \epsilon = 1 + 3x + 5x^2 + \epsilon$. Calculate the least-squares estimates of $\beta_0, \beta_1$ and $\beta_2$ and the estimate of $\sigma^2$. Carry out the sequential $F$-tests testing first for the quadratic term and then, if necessary, testing for the linear term.

2. In C1 place the $x$ values $-3.0, -2.5, -2.0, ..., 2.5, 3.0$. In C2 store a sample of 13 from the error $\epsilon$ where $\epsilon$ is distributed $N(0, 2)$. Fit the model $y = 1 + 3\cos(x) + 5\sin(x) + \epsilon$. Calculate the least-squares estimates of $\beta_0, \beta_1$ and $\beta_2$ and the estimate of $\sigma^2$. Carry out the $F$-test for any effect due to $x$. Are the sequential $F$-tests meaningful here?

3. In C1 place the $x_1$ values $-3.0$, $-2.5$, $-2.0$, ..., $2.5$, $3.0$. In C2 store a sample of 13 from the error $\epsilon$ where $\epsilon$ is distributed $N(0,2)$. In C3 store the values of $x_2 = x^2$. In C4 store the values of $y = 1 + 3\cos(x) + 5\sin(x) + \epsilon$. Now fit the model $y = \beta_0 + \beta_1 x_1 + \beta_2 x_2 + \epsilon$ and plot the standardized residuals in a normal quantile plot and against each of the explanatory variables.

# Chapter 12

# One-way Analysis of Variance

**New Minitab commands discussed in this chapter**

aovoneway   onewayaov

This chapter deals with methods for making inferences about the relationship existing between a single numeric response variable and a single categorical explanatory variable. The basic inference methods are the one-way analysis of variance (ANOVA) and the comparison of means. There are two commands for carrying out a one-way analysis of variance, namely **aovoneway** and **onewayaov**. They differ in the way the data must be stored for the analysis and **onewayaov** has many more subcommands available.

We write the one-way ANOVA model as $x_{ij} = \mu_i + \epsilon_{ij}$ where $i = 1, \ldots, I$ indexes the levels of the categorical explanatory variable and $j = 1, \ldots, n_i$ indexes the individual observations at each level, $\mu_i$ is the mean response at the $ith$ level and the errors $\epsilon_{ij}$ are a sample from the $N(0, \sigma)$ distribution. Based on the observed $x_{ij}$ we want to make inferences about the unknown values of the parameters $\mu_1, \ldots, \mu_I, \sigma$.

## 12.1   The ONEWAYAOV Command

We analyze the data of Example 12.6 in IPS using the **onewayaov** command. For this example there are $I = 3$ levels corresponding to the values Basal, DRTA and Strat and $n_1 = n_2 = n_3 = 22$. Suppose that we have the values of the $x_{ij}$ in C1 and the corresponding values of the categorical explanatory variable in C2 where Basal is indicated by 1, DRTA by 2 and Strat by 3. Then the command

```
MTB > onewayaov c1 c2 c3 c4;
SUBC> gboxplot;
SUBC> gnormal;
SUBC> gvariable c2;
SUBC> fisher.
```

```
Analysis of Variance for C1
Source     DF      SS      MS      F      P
C2          2    20.58   10.29   1.13   0.329
Error      63   572.45    9.09
Total      65   593.03
```

```
                                Individual 95% CIs For Mean
                                Based on Pooled StDev
Level  N   Mean   StDev   -----+---------+---------+---------+-
1     22  10.500  2.972        (----------*----------)
2     22   9.727  2.694     (----------*----------)
3     22   9.136  3.342  (----------*----------)
                         -----+---------+---------+---------+-
Pooled StDev = 3.014           8.4       9.6      10.8      12.0
```

```
Fisher's pairwise comparisons
 Family error rate = 0.121
Individual error rate = 0.0500
Critical value = 1.998
Intervals for (column level mean) - (row level mean)
                       1              2
         2         -1.043
                    2.589
         3         -0.452         -1.225
                    3.180          2.407
```

carries out a one-way ANOVA for the data in C1, with the levels in C2, and puts the ordinary residuals in C3 and the fitted values in C4. Note that because we assume a constant standard deviation the ordinary residuals can be used in place of standardized residuals. In Figure 12.1 we have plotted side-by-side boxplots of the data by level using the **gboxplot** subcommand.

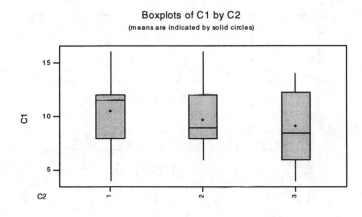

Figure 12.1: Boxplots for the data in Example 12.6 of IPS.

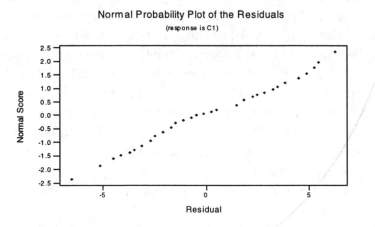

Figure 12.2: Normal quantile plot for the data in Example 12.6 of IPS after fitting a one-way ANOVA model.

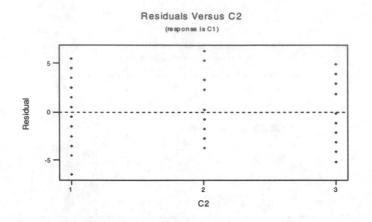

Figure 12.3: Plot of residuals against level in Example 12.6 of IPS after fitting a one-way ANOVA model.

In Figure 12.2 a normal quantile plot of the residuals is plotted using the **gnormal** subcommand and in Figure 12.3 we have plotted the residuals against level using the **gvariables** subcommand. None of the residual plots give us any reason to doubt the validity of the analysis. We could also obtain side-by-side dotplots of the data using the **gdotplot** subcommand, a histogram of the residuals using the **ghistogram** subcommand, a plot of the residuals against observation order using the **gorder** subcommand and a plot of the residuals against the fitted values using the **gfits** subcommand. Recall that these higher resolution plots may not be available if you are using an earlier version of Minitab but these plots can all be obtained using the saved residuals and fits. The $F$-test in the ANOVA table with a $P$-value of .329 indicates that the null hypothesis in the hypothesis testing problem $H_0 : \mu_1 = \mu_2 = \mu_3$ versus $H_0 : \mu_1 \neq \mu_2$ or $\mu_1 \neq \mu_3$ would not be rejected. Also the estimate of $\sigma$ is given by $s = 3.014$ and 95% confidence intervals are provided for the individual $\mu_i$.

The **fisher** subcommand gives confidence intervals for the differences between the means using

$$\bar{x}_i - \bar{x}_j \pm s\sqrt{\frac{1}{n_i} + \frac{1}{n_j}}t^*$$

where $s$ is the pooled standard deviation and $t^*$ is the .975 percentile of the Student distribution with the error degrees of freedom. Note that with an individual 95% confidence interval the probability of not covering the true difference – i.e. the *individual error rate* – is .05 but the probability of at least one of these three not covering the difference – i.e. the *family error rate* – is .121. If you want a more conservative family error rate, then specify a lower individual error rate on the same line as the **fisher** subcommand. For example an individual error rate of .02 specifies a family error rate of .0516 in this example. Also available for multiple comparisons are the **tukey**, **dunnett** and **mcb** subcommands and we refer the reader to **help** for details on these. Inferences for other contrasts of means can be carried out straightforwardly using the values of the estimates of the means and standard deviation provided by **onewayaov** and various Minitab commands.

In Exercise 12.5 we indicate how the power of the $F$-test can be approximated. As noted in IPS this involves the *noncentral F*-distribution.

In the window environment a one-way ANOVA can be carried out using Stat ▶ ANOVA ▶ One-way and filling in the dialog boxes appropriately. Also available are analysis of means (ANOM) plots via Stat ▶ ANOVA ▶ Analysis of Means (see **help** for details on these) and plots of the means with error bars (± one standard error of the observations at a level) via Stat ▶ ANOVA ▶ Interval Plots. Also we can plot the means joined by lines using Stat ▶ ANOVA ▶ Main Effects plots as in Figure 12.4. The dotted line is the grand mean. Power calculations can be carried out using Stat ▶ Power and Sample Size ▶ One-way ANOVA and filling in the dialog box appropriately.

## 12.2 The AOVONEWAY Command

The **aovoneway** command can be used for a one-way ANOVA when the data for each level is in a separate column. For example, suppose that the three samples for Example 12.6 in IPS are in columns C3-C5. Then the command

```
MTB > aovoneway c3-c5
```

produces the same ANOVA table and confidence intervals for the means as **onewayaov**. A major limitation of **aovoneway**, however, is that the only subcommands available are **gdotplot** and **gboxplot**. So if you have a worksheet with the samples for each level in columns it would seem better

Main Effects Plot - Data Means for C1

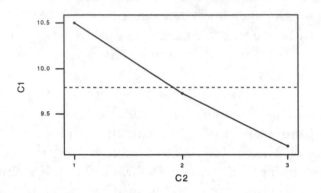

Figure 12.4: Main effects plot for Example 12.6 of IPS.

in general to use the **stack** command to place the data in one column and then use **onewayaov**.

In the window environment a one-way ANOVA can be carried out using Stat ▶ ANOVA ▶ One-way (Unstacked) and filling in the dialog box appropriately.

## 12.3   Exercises

When the data for an exercise come from an exercise in IPS, the IPS exercise number is given in parentheses ( ). All computations in these exercises are to be carried out using Minitab and the exercises are designed to ensure that you have a reasonable understanding of the Minitab material in this chapter. More generally you should be using Minitab to do all the computations and plotting required for the problems in IPS.

1. Generate a sample of 10 from each of the $N(\mu_i, \sigma)$ distributions for $i = 1, \ldots, 5$ where $\mu_1 = 1, \mu_2 = 1, \mu_3 = 1, \mu_4 = 1, \mu_5 = 2$ and $\sigma = 3$. Carry out a one-way ANOVA and plot a normal quantile plot of the residuals and the residuals against the explanatory variable. Compute .95 confidence intervals for the differences between the means. Compute an approximate set of .95 simultaneous confidence intervals for the

differences between the means.

2. Generate a sample of 10 from each of the $N(\mu_i, \sigma_i)$ distributions for $i = 1, \ldots, 5$ where $\mu_1 = 1, \mu_2 = 1, \mu_3 = 1, \mu_4 = 1, \mu_5 = 2$ and $\sigma_1 = \sigma_2 = \sigma_3 = \sigma_4 = 3$ and $\sigma_5 = 8$. Carry out a one-way ANOVA and plot a normal quantile plot of the residuals and the residuals against the explanatory variable. Compare the residual plots with those obtained in II.12.1.

3. If $X_1$ is distributed $Chisquare(k_1)$ independently of $X_2$ which is distributed $N(\delta, 1)$, then the random variable $Y = X_1 + X_2^2$ is distributed according to a *noncentral Chisquare*$(k + 1)$ distribution with noncentrality $\lambda = \delta^2$. Generate samples of $n = 1000$ from this distribution with $k = 2$ and $\lambda = 0, 1, 5, 10$. Plot histograms of these samples with the cut-points 0,1, ..., 200. Comment on the appearance of these histograms.

4. If $X_1$ is distributed *noncentral Chisquare*$(k_1)$ with non-centrality $\lambda$ independently of $X_2$ which is distributed $Chisquare(k_2)$, then the random variable

$$Y = \frac{X_1/k_1}{X_2/k_2}$$

is distributed according to a *noncentral* $F(k_1, k_2)$ distribution with non-centrality $\lambda$. Generate samples of $n = 1000$ from this distribution with $k_1 = 2, k_2 = 3$ and $\lambda = 0, 1, 5, 10$. Plot histograms of these samples with the cut-points 0,1, ..., 200. Comment on the appearance of these histograms.

5. As noted in IPS the $F$-statistic in a one-way ANOVA, when the standard deviation $\sigma$ is constant from one level to another, is distributed *noncentral* $F(k_1, k_2)$ with noncentrality $\lambda$ where $k_1 = I - 1$, $k_2 = n_1 + \cdots n_I - I$,

$$\lambda = \frac{\sum_{i=1}^{I} n_i \left(\mu_i - \bar{\mu}\right)^2}{\sigma^2}$$

and $\bar{\mu} = \sum_{i=1}^{I} n_i \mu_i / \sum_{i=1}^{I} n_i$. Using simulation approximate the power of the test in problem II.12.1 with level .05 and the values of the parameters specified.

# Chapter 13

# Two-way Analysis of Variance

**New Minitab commands discussed in this chapter**

**twowayaov**

This chapter deals with methods for making inferences about the relationship existing between a single numeric response variable and two categorical explanatory variables. The **twowayaov** command is used to carry out a two-way ANOVA.

We write the two-way ANOVA model as $x_{ijk} = \mu_{ij} + \epsilon_{ijk}$ where $i = 1, \ldots, I$ and $j = 1, \ldots, J$ index the levels of the categorical explanatory variables and $k = 1, \ldots, n_{ij}$ indexes the individual observations at each treatment (combination of levels), $\mu_{ij}$ is the mean response at the *ith* level and the *jth* level of the first and second explanatory variable respectively and the errors $\epsilon_{ijk}$ are a sample from the $N(0, \sigma)$ distribution. Based on the observed $x_{ijk}$ we want to make inferences about the unknown values of the parameters $\mu_{11}, \ldots, \mu_{IJ}, \sigma$.

## 13.1 The TWOWAYAOV Command

We consider a generated example where $I = J = 2, \mu_{11} = \mu_{21} = \mu_{12} = \mu_{22} = 1, \sigma = 2$ and $n_{11} = n_{21} = n_{12} = n_{22} = 5$. The $\epsilon_{ijk}$ are generated as a sample from the $N(0, \sigma)$ distribution and then we put $x_{ijk} = \mu_{ij} + \epsilon_{ijk}$ for $i = 1, \ldots, I$ and $j = 1, \ldots, J$ and $k = 1, \ldots, n_{ij}$. Note that the **twowayaov** command requires balanced data; i.e. all the $n_{ij}$ must be equal. Now we pretend that we don't know the values of the parameters and carry out a two-way analysis of variance. If the $x_{ijk}$ are in C1, the values of $i$ in C2 and the values of $j$ in C3, then the command

```
MTB > twowayaov c1 c2 c3 c4 c5;
SUBC> gnormal;
SUBC> gvariable c2 c3;
SUBC> means c2 c3.
```

```
Analysis of Variance for C1
Source      DF      SS      MS      F      P
C2           1    0.39    0.39   0.07  0.790
C3           1    3.43    3.43   0.65  0.432
Interaction  1    4.01    4.01   0.76  0.396
Error       16   84.44    5.28
Total       19   92.26
```

```
                Individual 95% CI
C2    Mean  -+---------+---------+---------+---------+
1     1.49  (------------------*------------------)
2     1.77       (-----------------*-----------------)
            -+---------+---------+---------+---------+
          0.00      0.80      1.60      2.40      3.20
```

```
                Individual 95% CI
C3    Mean  ----+---------+---------+---------+-------
1     2.05       (--------------*---------------)
2     1.22  (--------------*---------------)
            ----+---------+---------+---------+-------
          0.00      1.00      2.00      3.00
```

prints out the two-way ANOVA table and stores the residuals in C4 and the fitted values in C5. We see from this that the null hypothesis of no interaction is not rejected ($P$-value $= .396$) and neither is the null hypothesis of no effect due to the C2 factor ($P$-value $= .790$) nor the null hypothesis of no effect due to factor C3 ($P$-value $= .432$). Also using the **gnormal** subcommand a normal quantile plot of the residuals is plotted while the **gvariables** subcommand causes a plot of the residuals against each of the factors C2 and C3. We haven't reproduced those plots here but they look acceptable in this example as we might expect they would. The **ghistogram**, **gfits** and

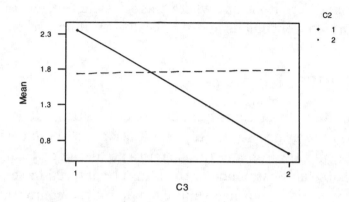

Figure 13.1: Plot of cell means in two-way ANOVA simulated example.

**gorder** subcommands are also available for a histogram of the residuals, the residuals against the fitted values and the residuals against observation order respectively. The **means** subcommand causes the estimates of marginal means for each level of C2 and C3 to be printed together with 95% confidence intervals. Note that it is only appropriate to look at these marginal means if we do not reject the null hypothesis of no interaction between the factors. If we conclude that there is an interaction then we must look at the individual $IJ$ cell means to determine where the interaction occurs. A plot of these cell means is often useful in this regard.

If we want to fit the model without any interaction, supposing we know this to be true, then the **additive** subcommand is available to do this. This is acceptable only in very rare circumstances.

In the window environment a two-way ANOVA can be carried out using Stat ► ANOVA ► Two-way and filling in the dialog boxes appropriately. Also available are analysis of means (ANOM) plots via Stat ► ANOVA ► Analysis of Means. Also we can plot the marginal means joined by lines using Stat ► ANOVA ► Main Effects Plot and plot the cell means joined by lines using Stat ► ANOVA ► Interaction Plot as in Figure 13.1. Note that while the plot seems to indicate an interaction this is not confirmed by the statistical test. Power calculations can be carried out using Stat ► Power and Sample Size ► 2-Level Factorial Design and filling in the dialog box appropriately. In Version 12 there are features available for analyzing unbalanced data and for situations where there are more than two factors where some factors are continuous and some categorical and so on.

## 13.2 Exercises

When the data for an exercise come from an exercise in IPS, the IPS exercise number is given in parentheses ( ). All computations in these exercises are to be carried out using Minitab and the exercises are designed to ensure that you have a reasonable understanding of the Minitab material in this chapter. More generally you should be using Minitab to do all the computations and plotting required for the problems in IPS.

1. Suppose $I = J = 2, \mu_{11} = \mu_{21} = 1$ and $\mu_{12} = \mu_{22} = 2, \sigma = 2$ and $n_{11} = n_{21} = n_{12} = n_{22} = 10$. Generate the data for this situation and carry out a two-way analysis. Plot the cell means (an interaction effect plot). Do your conclusions agree with what you know to be true?

2. Suppose $I = J = 2, \mu_{11} = \mu_{21} = 1$ and $\mu_{12} = 3, \mu_{22} = 2, \sigma = 2$ and $n_{11} = n_{21} = n_{12} = n_{22} = 10$. Generate the data for this situation and carry out a two-way analysis. Plot the cell means ( an interaction effect plot). Do your conclusions agree with what you know to be true?

3. Suppose $I = J = 2, \mu_{11} = \mu_{21} = 1$ and $\mu_{12} = \mu_{22} = 2, \sigma = 2$ and $n_{11} = n_{21} = n_{12} = n_{22} = 10$. Generate the data for this situation and carry out a two-way analysis. Form 95% confidence intervals for the marginal means. Repeat your analysis using the **additive** subcommand and compare the confidence intervals. Can you explain your results?

# Chapter 14

# Nonparametric Tests

**New Minitab commands discussed in this chapter**

friedman   kruskal-wallis   mann-whitney   winterval   wtest

This chapter deals with inference methods that do not depend upon the assumption of normality. These methods are sometimes called *nonparametric* or *distribution free* methods. Recall that we discussed a distribution free method in II.7.4 where we presented the **sinterval** and **stest** commands for the sign confidence interval and sign test for the median. Recall also the **rank** command in I.11.9 which can be used to compute the ranks of a data set.

## 14.1  The MANN-WHITNEY Command

The Mann-Whitney test for a difference between the locations of two distributions is equivalent to the Wilcoxon rank sum test in the following sense. Suppose that we have two independent samples $y_{11}, \ldots, y_{1n_1}$ and $y_{21}, \ldots, y_{2n_2}$ from two distributions that differ at most in their locations as represented by their medians. The Mann-Whitney statistic $U$ is the number of pairs $(y_{1i}, y_{2j})$ where $y_{1i} > y_{2j}$ while the Wilcoxon rank sum test statistic $W$ is the sum of the ranks from the first sample when the ranks are computed for the two samples considered as one sample. Then it can be shown that $W = U + n_1(n_{1+1})/2$.

For Example 14.1 of IPS  we store the four values 166.7, 172.2, 165.0, 176.9 of sample 1 in C1 and the four values 158.6, 176.4, 153.1, 156.0 of sample 2 in C2. Then the command

```
MTB > mann-whitney 90 c1 c2;
SUBC> alternative 1.
Mann-Whitney Confidence Interval and Test
C1 N = 4 Median = 169.45
C2 N = 4 Median = 157.30
Point estimate for ETA1-ETA2 is 11.30
93.9 Percent CI for ETA1-ETA2 is (-9.70,20.90)
W = 23.0
Test of ETA1 = ETA2 vs ETA1 > ETA2 is significant at 0.0970
Cannot reject at alpha = 0.05
```

indicates that the test of $H_0$ : the medians of the two distributions are identical versus $H_a$ : the median of the first distribution is greater than the median of the second gives a $P$-value of .0970. Also an estimate of 11.3 is produced for the difference in the medians and we asked for a 90% confidence interval for this difference by including 90 on the command line. If this value is left out a default 95% confidence interval is computed. Note that exact confidences cannot be attained due to the discrete distribution followed by the statistic $U$. Also available are the one-sided test of $H_0$ : the medians of the two distributions are identical versus $H_a$ : the median of the first distribution is smaller than the median of the second, using the subcommand **alternative** $-1$ and the two-sided test is obtained if no **alternative** subcommand is employed. The Mann-Whitney test requires the assumption that the two distributions we are sampling from have the same form.

In the window environment the Mann-Whitney test and confidence interval are available via S̲tat ▶ N̲onparametrics ▶ M̲ann-Whitney and filling in the dialog box appropriately.

## 14.2   WTEST and WINTERVAL Commands

The Wilcoxon signed rank test and confidence interval are used for inferences about the median of a distribution. The Wilcoxon procedures are based on ranks which is not the case for the sign procedures discussed in II.7.4. Consider the data of Example 14.8 in IPS where the differences between two scores has been recorded as .37, −.23, .66, −.08, −.17 in C1. Then the command

```
MTB > wtest c1;
SUBC> alternative 1.
Test of median = 0.000000 versus median > 0.000000
            N for    Wilcoxon             Estimated
       N    Test     Statistic      P     Median
C1     5    5              9.0    0.394    0.1000
```

gives the $P$-value .394 for testing $H_0$ : the median of the difference is 0 versus $H_a$ : the median of the difference is greater than 0. The **alternative** sub-command works just as it does with the other testing commands in Minitab. The general syntax of the **wtest** command is

**wtest** V $E_1$

where V is the hypothesized value of the median, with 0 being the default value, and $E_1$ is the column containing the data. For confidence intervals the command

```
MTB > winterval 90 c1
            Estimated    Achieved
       N    Median       Confidence    Confidence Interval
C1     5    0.100        89.4          ( -0.200, 0.515)
```

computes a 90% confidence interval for the median. Note that the Wilcoxon signed rank procedures for the median require an assumption that the response values (in this case the difference) come from a distribution symmetric about its median.

In the window environment the Wilcoxon signed rank test and confidence interval are available via Stat ▶ Nonparametrics ▶ 1-Sample Wilcoxon and filling in the dialog box appropriately.

# 14.3   The KRUSKAL-WALLIS Command

The Kruskal-Wallis test is the analog of the one-way ANOVA in the nonparametric setting. Suppose the data for Example 14.13 in IPS are in C1 and C2 where C1 contains the corn yield in bushels per acre and C2 is number of weeds per meter. Then the command

```
MTB > kruskal-wallis c1 c2
Kruskal-Wallis Test on C1
C2        N    Median   Ave Rank       Z
1         4    169.4        13.1    2.24
2         4    163.6         8.4   -0.06
3         4    157.3         6.2   -1.09
9         4    162.6         6.2   -1.09
Overall 16 8.5
H = 5.56 DF = 3 P = 0.135
H = 5.57 DF = 3 P = 0.134 (adjusted for ties)
* NOTE * One or more small samples
```

gives a $P$-value of .135 for testing $H_0$ : each sample comes from the same distribution versus $H_a$ : at least two of the samples come from different distributions. Note that the validity of the Kruskal-Wallis test relies on the assumption that the distributions being sampled from all have the same form.

In the window environment the Kruskal-Wallis test is available via Stat ▶ Nonparametrics ▶ Kruskal-Wallis and filling in the dialog box appropriately.

## 14.4   The FRIEDMAN Command

The Friedman test is the analog of the two-way ANOVA test when we assume there is no interaction between the factors. Basically these situations arise in what are called *matched samples*; e.g. $a$ persons are asked to score $b$ brands of chocolate chip cookies for flavor and we want to test for a difference amongst the brands. In this case the factor "person" is referred to as the *blocking factor* and we are not interested in the effect due to this factor. A blocking factor is introduced into a study to reduce the contribution of error to the experimental results and it is assumed to have an effect or we wouldn't consider it. Suppose in C11 we have the scores (in this case one of the values 0, 1, ..., 10 reflecting how "tasty" the cookie is) given to $b = 6$ brands of cookies – the *treatment* – by $a = 10$ people, in C12 we have the brand factor and in C13 we have the block factor – an indicator for individual. Then the command

```
MTB > friedman c11 c12 c13
Friedman test for C11 by C12 blocked by C13
S = 9.49 DF = 5 P = 0.091
```

```
S = 9.85 DF = 5 P = 0.080 (adjusted for ties)
                   Est      Sum of
C12     N        Median     Ranks
1      10         3.854      32.5
2      10         6.937      46.0
3      10         3.938      32.5
4      10         4.771      31.0
5      10         3.188      24.5
6      10         6.937      43.5
Grand median = 4.937
```

gives a $P$-value of .091 for $H_0$ : there is no difference among the brands versus $H_a$ : there is a difference among the brands, so we do not reject the null hypothesis at the .05 level. Residuals and fitted values based on medians can also be computed and stored and we refer the reader to **help** for more details on this. The Friedman test is valid under the assumption that all the distributions for each level of the treatment are of the same form.

In the window environment the Friedman test is available via S̲tat ▶ N̲onparametrics ▶ K̲ruskal-Wallis and filling in the dialog box appropriately.

## 14.5 Exercises

When the data for an exercise come from an exercise in IPS, the IPS exercise number is given in parentheses ( ). All computations in these exercises are to be carried out using Minitab and the exercises are designed to ensure that you have a reasonable understanding of the Minitab material in this chapter. More generally you should be using Minitab to do all the computations and plotting required for the problems in IPS.

1. Generate a sample of $n = 10$ from the $N(0, 1)$ distribution and compute the $P$-value for testing $H_0$ : the median is 0 versus $H_a$ : the median is not 0, using the $t$-test and the Wilcoxon signed rank test. Compare the $P$-values. Repeat this with $n = 100$.

2. Generate a sample of $n = 10$ from the $N(0, 1)$ distribution and compute 95% confidence intervals for the median, using the $t$-confidence interval and the Wilcoxon signed rank confidence intervals. Compare the lengths of the confidence intervals. Repeat this with $n = 100$.

3. Generate two samples of $n = 10$ from the *Student*$(1)$ distribution and to the second sample add 1. Then test $H_0$ : the medians of the two distributions are identical versus $H_a$ : the medians are not equal using the two sample $t$-test and using the Mann-Whitney test. Compare the results.

4. Generate a sample of 10 from each of the $N(1,2), N(2,2)$ and $N(3,1)$ distributions. Test for a difference amongst the distributions using a one-way ANOVA and using the Kruskal-Wallis test. Compare the results.

5. Generate 10 scores for 10 brands from the $N(\mu_{ij}, \sigma)$ distributions for $i = 1, 2$ and $j = 1, 2$ where $\mu_{11} = \mu_{21} = 1$ and $\mu_{12} = \mu_{22} = 2$ and treat each test for no effect due to brand using a two-way ANOVA with the assumption of no interaction and also using the Friedman test. Compare the results.

# Chapter 15

# Logistic Regression

This chapter deals with the *logistic regression model*. This model arises when the response variable $y$ is binary, i.e. takes only two values, and we have a number of explanatory variables $x_1, \ldots, x_k$. Version 12 of Minitab has commands for carrying out logistic regression.

## 15.1   The Logistic Regression Model

The regression techniques discussed in Chapters 10 and 11 require that the response variable $y$ be a continuous variable. In many contexts, however, the response is discrete and in fact binary,; i.e. taking the values 0 and 1. Let $p$ denote the probability of a 1. Then this probability is related to the values of the explanatory variables $x_1, \ldots, x_k$. We cannot, however, write this as $p = \beta_0 + \beta_1 x_1 + \ldots + \beta_k x_k$ because the right-hand side is not constrained to lie in the interval [0,1] which it must if it is to represent a probability. One solution to this problem is to employ the *logit link function* which is given by

$$\ln\left(\frac{p}{1-p}\right) = \beta_0 + \beta_1 x_1 + \cdots + \beta_k x_k$$

and this leads to the equations

$$\frac{p}{1-p} = \exp\{\beta_0 + \beta_1 x_1 + \cdots + \beta_k x_k\}$$

and

$$p = \frac{\exp\{\beta_0 + \beta_1 x_1 + \cdots + \beta_k x_k\}}{1 + \exp\{\beta_0 + \beta_1 x_1 + \cdots + \beta_k x_k\}}$$

for the *odds* $p/(1-p)$ and probability $p$ respectively. The right-hand side of the equation for $p$ is now always between 0 and 1. Note that logistic regression is based on an ordinary regression relation between the logarithm of the odds in favor of the event occurring at a particular setting of the explanatory variables and the values of the explanatory variables $x_1, \ldots, x_k$. The quantity $\ln(p/(1-p))$ is referred to as the *log odds*.

The procedure for estimating the coefficients $\beta_0, \beta_1, \ldots, \beta_k$ using this relation and carrying out tests of significance on these values is known as *logistic regression*. Typically more sophisticated statistical methods than least squares are needed for fitting and inference in this context and we rely on software such as Minitab to carry out the necessary computations.

There are also other link functions that are often used and these are available in Minitab. In particular the *probit link function* is given by

$$\Phi^{-1}(p) = \beta_0 + \beta_1 x_1 + \cdots + \beta_k x_k$$

where $\Phi$ is the cumulative distribution function of the $N(0,1)$ distribution and this leads to the relation

$$p = \Phi(\beta_0 + \beta_1 x_1 + \cdots + \beta_k x_k)$$

which is also always between 0 and 1. Choice of the link function can be made via a variety of goodness-of-fit tests available in Minitab but we restrict our attention here to the logit link function.

## 15.2 Example

Suppose now that we have the following 10 observations in columns C1-C3

| Row | C1 | C2 | C3 |
|-----|----|----|----|
| 1 | 0 | -0.65917 | 0.43450 |
| 2 | 0 | 0.69408 | 0.48175 |
| 3 | 1 | -0.28772 | 0.08279 |
| 4 | 1 | 0.76911 | 0.59153 |
| 5 | 1 | 1.44037 | 2.07466 |
| 6 | 0 | 0.52674 | 0.27745 |
| 7 | 1 | 0.38593 | 0.14894 |
| 8 | 1 | -0.00027 | 0.00000 |
| 9 | 0 | 1.15681 | 1.33822 |
| 10 | 1 | 0.60793 | 0.36958 |

where the response $y$ is in C1, $x_1$ is in C2 and $x_2$ is in C3 and note that $x_2 = x_1^2$. Then we want to fit the model

$$\ln\left(\frac{p}{1-p}\right) = \beta_0 + \beta_1 x_1 + \beta_2 x_2$$

and conduct statistical inference concerning the parameters of the model.

Fitting and inference is carried out in Minitab using Stat ▶ Regression ▶ Binary Logistic Regression and filling in the dialog box so that the Response box contains C1, the Model box contains C2 and C3 and under Results the radio button Response information, regression table, etc. is filled in. The choice under Results is not necessary as this only controls the amount of output but the default output is more extensive and we chose to limit this. The output

```
Link Function:  Logit
Response Information

Variable Value Count
C1        1        6 (Event)
          0        4
       Total      10

Logistic Regression Table
```

| Predictor | Coef | StDev | Z | P | Odds Ratio | 95% CI Lower | Upper |
|-----------|------|-------|---|---|------------|-------|-------|
| Constant | 0.5228 | 0.9031 | 0.58 | 0.563 | | | |
| C2 | 0.740 | 1.605 | 0.46 | 0.645 | 2.10 | 0.09 | 48.71 |
| C3 | -0.780 | 1.584 | -0.49 | 0.623 | 0.46 | 0.02 | 10.23 |

```
Log-Likelihood = -6.598
Test that all slopes are zero:   G = 0.265, DF = 2,
                                             P-Value = 0.876
```

gives estimates of the coefficients and their standard errors and gives the $P$-value for $H_0 : \beta_0 = 0$ versus $H_a : \beta_0 \neq 0$ as 0.563, the $P$-value for $H_0 : \beta_1 = 0$ versus $H_a : \beta_1 \neq 0$ as 0.643 and the $P$-value for $H_0 : \beta_2 = 0$ versus $H_a : \beta_2 \neq 0$ as 0.623. Further the test of $H_0 : \beta_1 = \beta_2 = 0$ versus $H_a : \beta_1 \neq 0$ or $\beta_2 \neq 0$ has $P$-value .876. In this example there is no evidence of any nonzero coefficients. Note that when $\beta_0 = \beta_1 = \beta_2 = 0$, $p = .5$.

Also provided in the output is the estimate 2.10 for the odds ratio for $x_1$ (C2) and a 95% confidence interval $(.09, 48.71)$ for the true value. The odds ratio for $x_1$ is given by $\exp(\beta_1)$ which is the ratio of the odds at $x_1 + 1$ to the odds at $x_1$ when $x_2$ is held fixed or when $\beta_2 = 0$. Since there is evidence that $\beta_2 = 0$ ($P$-value $= .623$) the odds ratio has a direct interpretation here. Note, however, that if this wasn't the case then the odds ratio would not have such an interpretation as it doesn't makes sense for $x_2$ to be held fixed when $x_1$ changes in this example as they are not independent variables. Similar comments apply to the estimate 0.46 for the odds ratio for $x_2$ (C3) and the 95% confidence interval $(.02, 10.23)$ for the true value of this quantity.

Many other aspects of fitting logistic regression models are available in Minitab and we refer the reader to **help** for a discussion of these. Also available in Minitab are *ordinal logistic regression*, when the response takes more than two values and these are ordered, and *nominal logistic regression*, when the response takes more than two values and these are unordered. These can be accessed via S̲tat ▶ R̲egression ▶ Ordinal L̲ogistic Regression and S̲tat ▶ R̲egression ▶ Nominal L̲ogistic Regression respectively.

## 15.3   Exercises

When the data for an exercise come from an exercise in IPS, the IPS exercise number is given in parentheses ( ). All computations in these exercises are to be carried out using Minitab and the exercises are designed to ensure that you have a reasonable understanding of the Minitab material in this chapter. More generally you should be using Minitab to do all the computations and plotting required for the problems in IPS.

1. Generate a sample of 20 from the *Bernoulli*$(.25)$ distribution. Pretend-

ing that we don't know $p$ compute a 95% confidence interval for this quantity. Using this confidence interval form 95% confidence intervals for the odds and the log odds.

2. Let $x$ take the values $-1, -.5, 0, .5$ and $1$. Plot the log odds

$$\ln\left(\frac{p}{1-p}\right) = \beta_0 + \beta_1 x$$

against $x$ when $\beta_0 = 1$ and $\beta_1 = 2$. Plot the odds and the probability $p$ against $x$.

3. Let $x$ take the values $-1, -.5, 0, .5$ and $1$. At each of these values generate a sample of 4 values from the $Bernoulli(p_x)$ distribution where

$$p_x = \frac{\exp\{1 + 2x\}}{1 + \exp\{1 + 2x\}}$$

and let these values be the $y$ response values. Carry out a logistic regression analysis of this data using the model .

$$\ln\left(\frac{p_x}{1-p_x}\right) = \beta_0 + \beta_1 x$$

Compute a 95% confidence interval for $\beta_1$ and determine if it contains the true value. Similarly form a 95% confidence interval for the odds ratio when $x$ increases by 1 unit and determine if it contains the true value.

4. Let $x$ take the values $-1, -.5, 0, .5$ and $1$. At each of these values generate a sample of 4 values from the $Bernoulli(p_x)$ distribution where

$$p_x = \frac{\exp\{1 + 2x\}}{1 + \exp\{1 + 2x\}}$$

and let these values be the $y$ response values. Carry out a logistic regression analysis of this data using the model

$$\ln\left(\frac{p_x}{1-p_x}\right) = \beta_0 + \beta_1 x + \beta_2 x^2$$

Test the null hypothesis $H_0 : \beta_2 = 0$ versus $H_a : \beta_2 \neq 0$. Form a 95% confidence interval for the odds ratio for $x$. Does it make sense to make an inference about this quantity in this example? Why or why not?

5. Let $x$ take the values $-1$, $-.5$, $0$, $.5$ and $1$. At each of these values generate a sample of 4 values from the $Bernoulli(.5)$ distribution. Carry out a logistic regression analysis of this data using the model

$$\ln\left(\frac{p_x}{1 - p_x}\right) = \beta_0 + \beta_1 x + \beta_2 x^2$$

Test the null hypothesis $H_0 : \beta_1 = \beta_2 = 0$ versus $H_a : \beta_1 \neq 0$ or $\beta_2 \neq 0$.

# Appendix A

# Projects

The basic structural component of Minitab is the worksheet. When working on a project it may make sense to have your data in several worksheets so that similar variables are grouped together. Also you may wish to save high resolution plots associated with the worksheets so that everything can be obtained via a single reference. In Version 12 worksheets and graphs can be grouped together into *projects*. Projects are given names and are stored in a file with the supplied name and the file extension .mpj.

To open a new project use File ▶ New and then choose Minitab Project and click OK. If you want to open a previously saved project then use File ▶ Open Project and choose the relevant project from the list. To save a project use File ▶ Save Project if the project already has a name (or you wish to use the default of minitab) or File ▶ Save Project As if you wish to give the project a name. Not only are the contents of all worksheets and graphs saved but the contents of the History window is saved as well and is available when the project is reopened. You can also supply a description of the project using File ▶ Project Description and filling in the dialog box. Note that a description of a worksheet can also be saved using Editor ▶ Worksheet Description. When you attempt to open a new project or exit Minitab you will be asked if you wish to save the contents of the current project.

Now suppose that in the project evans we have a single worksheet containing 100 numeric values in each of C1 and C2 and have produced a scatter plot of C2 against C1. We now open a new worksheet using File ▶ New and then choose Minitab Worksheet and click OK. There are now two worksheets associated with the project called Worksheet1 and Worksheet2. Suppose that we also place 100 numeric values in C1 and C2 in Worksheet 2 and again plot

C2 against C1. We now also have two plots associated with the project **evans** called Worksheet 1: Plot C2*C1 and Worksheet 2: Plot C2*C1. These will all appear as individual windows on your screen, perhaps with some hidden, and any one in particular can be made active by clicking in that window or by clicking on the relevant entry in the list obtained when you use Window. You can also save individual worksheets in the project to files outside the project when a particular worksheet is active using the **save** command or File ▶ Save Current Worksheet As. Similarly, when a graph window is active a graph in the project can be saved to a file outside the project using File ▶ Save Graph As.

With multiple worksheets in a project it is easy to move data between worksheets using cut, copy and paste operations. For example, suppose that we want to copy C1 and C2 of Worksheet 1 into C3 and C4 of Worksheet 2. Then, with Worksheet 1 active, highlight the entries in C1 and C2, use Edit ▶ Copy Cells, make Worksheet 2 active, click in the first cell of C3 and then use Edit ▶ Paste Cells.

It is possible to see what a project contains without opening it. To do this use File ▶ Open Project, click on the project to be previewed and then click on the Preview button. Similarly worksheets can be previewed using File ▶ Open Worksheet, clicking on the worksheet to be previewed and then clicking on the Preview button.

# Appendix B

# Mathematical and Statistical Functions in Minitab

The explanation of the behavior of the various functions corresponds to Version 12. If a function is available only in Version 12 then this is noted.

## B.1 Mathematical Functions

Here is a list and description of the mathematical and statistical functions available in Minitab. All of these functions operate on each element of a column and return a column of the same length. Let $(x_1, \ldots, x_n)$ denote a column of length $n$. These functions can be applied only to numerical variables.

**absolute** - computes the absolute value, $(|x_1|, \ldots, |x_n|)$.

**antilog** - computes the inverse of the base 10 logarithm, $(10^{x_1}, \ldots, 10^{x_n})$.

**acos** - computes the inverse cosine function, $(\arccos(x_1), \ldots, \arccos(x_n))$.

**asin** - computes the inverse sine function, $(\arcsin(x_1), \ldots, \arcsin(x_n))$.

**atan** - computes the inverse tan function, $(\arctan(x_1), \ldots, \arctan(x_n))$ ..

**cos** - computes the cosine function when angle is given in radians,
$(\cos(x_1), \ldots, \cos(x_n))$.

**ceiling** - computes the smallest integer bigger than a number,
$(\lceil x_1 \rceil, \ldots, \lceil x_n \rceil)$, (only in Version 12).

**degrees** - computes the degree measurement of an angle given in radians,
(only in Version 12).

**exponentiate** - computes the exponential function, $(e^{x_1}, \ldots e^{x_n})$.

**floor** - computes the greatest integer smaller than a number,

$(\lfloor x_1 \rfloor, \ldots, \lfloor x_n \rfloor)$ (only in Version 12).

**gamma** - computes the gamma function, $(\Gamma(x_1), \ldots, \Gamma(x_n))$; note that for nonnegative integer $x$, $\Gamma(x+1) = x!$ (only in Version 12).

**lag** - computes the column $(*, x_1, \ldots, x_{n-1})$.

**log-gamma** - computes the log-gamma function, $(\ln \Gamma(x_1), \ldots, \ln \Gamma(x_n))$; note that for nonnegative integer $x$, $\ln \Gamma(x+1) = \sum_{i=1}^{x} \ln(i)$ (only in Version 12).

**loge** - computes the natural logarithm function, $(\ln(x_1), \ldots, \ln(x_n))$.

**logten** - computes the base 10 logarithm function, $(\log_{10}(x_1), \ldots, \log_{10}(x_n))$.

**nscore** - computes the normal scores function; see **help**.

**parsums** - computes the column of partial sums,

$(x_1, x_1 + x_2, \ldots, x_1 + \cdots x_n)$.

**parproducts** - computes the column of partial products,

$(x_1, x_1 x_2, \ldots, x_1 \cdots \cdots x_n)$.

**radians** - computes the radian measurement of an angle given in degrees (only in Version 12).

**rank** - computes the ranks of the column entries, $(r_1, \ldots r_n)$.

**round** - computes the nearest integer function $i(x)$ with rounding up at .5,

$(i(x_1), \ldots, i(x_n))$; see **help** for more details on this function.

**signs** - computes the sign function

$$s(x) = \begin{cases} -1 & if \quad x < 0 \\ 0 & if \quad x = 0 \\ 1 & if \quad x > 0 \end{cases}$$

$(s(x_1), \ldots, s(x_n))$.

**sin** - computes the sine function when the angle is given in radians,

$(\sin(x_1), \ldots, \sin(x_n))$.

**sort** - computes the column consisting of the sorted (ascending) column entries, $(x_{(1)}, \ldots, x_{(n)})$.

**sqrt** - computes the square root function, $(\sqrt{x_1}, \ldots, \sqrt{x_n})$.

**tan** - computes the tan function when the angle is given in radians,

$(\tan(x_1), \ldots, \tan(x_n))$.

# B.2  Column Statistics

Let $(x_1, \ldots, x_n)$ denote a column of length $n$. Output is written on the screen or in the Session window and can be assigned to a constant. The general syntax for column statistic commands is

**column statistic name($E_1$)**

where the operation is carried out on the entries in column $E_1$ and output is written to the screen unless it is assigned to a constant using the **let** command.

**max** - computes the maximum of a column, $x_{(n)}$.
**mean** - computes the mean of a column, $\bar{x} = (x_1 + \cdots x_n)/n$.
**median** - computes the median of a column (see Chapter 1).
**min** - computes the minimum of a column, $x_{(1)}$.
**n** - computes the number of nonmissing values in the column.
**nmiss** - computes the number of missing values in the column.
**range** - computes the difference between the smallest and largest value in a column, $x_{(n)} - x_{(1)}$.
**ssq** - computes the sum of squares of a column, $x_1^2 + \cdots + x_n^2$.
**stdev** - computes the standard deviation of a column,

$$s = \sqrt{\frac{1}{n-1}\left[(x_1 - \bar{x})^2 + \cdots + (x_n - \bar{x})^2\right]}$$

**sum** - computes the sum of the column entries, $x_1 + \cdots x_n$.

# B.3  Row Statistics

Let $(x_1, \ldots, x_n)$ denote a row of length $n$. The general syntax is

**row statistic name $E_1 \ldots E_m$ $E_{m+1}$**

where the operations are carried out on the rows in columns $E_1, \ldots, E_m$ and the output is placed in column $E_{m+1}$.

**rmax** - computes the maximum of a row, $x_{(n)}$.
**rmean** - computes the mean of a row, $\bar{x} = (x_1 + \cdots x_n)/n$.
**rmiss** - computes the number of missing values in the row.
**rn** - computes the number of nonmissing values in the row.

**rrange** - computes the difference between the smallest and largest value in a row, $x_{(n)} - x_{(1)}$.

**rssq** - computes the sum of squares of a row, $x_1^2 + \cdots + x_n^2$.

**rstdev** - computes the standard deviation of a row,

$$s = \sqrt{\frac{1}{n-1}\left[(x_1 - \bar{x})^2 + \cdots + (x_n - \bar{x})^2\right]}$$

**rsum** - computes the sum of the row entries, $x_1 + \cdots x_n$.

# Appendix C

# Macros and Execs

We can store Minitab commands in files that can be called to execute these commands without having to type them. Also we can program Minitab to carry out iterative calculations. These aspects involve us in a discussion of *macros* and *execs* in Minitab. We present a very brief overview of this topic and refer the reader to reference [2] in Appendix F for a more extensive discussion. There are two types of macros, *global* and *local*. If you are using Version 12 then you need only read about global and local macros; users of earlier versions need only read about execs. The capabilities of execs are covered by macros.

Note that, because it is possible to write a macro or exec that will loop endlessly, it is important to know how to stop the execution if you feel it is running too long. To do this, simultaneously press the Control and Break keys.

If the macro processor in Minitab finds an error in a macro, then this is indicated in the Session window by **ERROR**. If there is an error in a Minitab command in the macro, then this is indicated by *ERROR*. Minitab also provides a message that attempts to diagnose what has caused the problem.

## C.1   Global Macros

A global macro is a set of commands in a file with the structure

> gmacro
> *template*

*body*
endmacro

where *template* is a name for the macro, consisting of any characters but starting with a letter, and *body* is a set of Minitab commands, macro statements or other macro names. In general it is good form to use the file name for *template* but this is not necessary. If the file has the extension .mac, then only the file name, immediately preceded by %, need be used to invoke it. Otherwise the full file name and extension must be used and again immediately preceded by %. Also the full path name must be used unless the file is in the default Minitab directory or in the **macros** subdirectory of the default Minitab directory. The statements **gmacro** and **endmacro** must always start and end the file respectively.

Suppose that we have placed the following statements

```
gmacro
generate
note This macro generates 50 samples of size 20 from the Uniform(0,1).
do k1=1:50
random 20 c1;
uniform 0 1.
let c2(k1)=mean(c1)
enddo
endmacro
```

in a file called `generate.txt`. Then the macro is invoked via the Minitab command

```
MTB > %generate.txt
```

and causes 50 samples of size 20 to be generated from the uniform distribution on the interval $(0, 1)$, with each sample stored in C1 overwriting the preceding one, and causes the sample mean to be computed for each of these and to be stored in the corresponding element of C2. Finally a histogram is produced of these 50 means. Clearly this is a much more powerful method for carrying out simulations than the one we discussed in II.6 as that was in essence limited by the size of the worksheet. Note the use of the **do, enddo** statements to perform the calculations iteratively. Also the **note** command is used to display the text following it, on the same line, in the Session window. Otherwise there is nothing printed in the Session window beyond the output from any commands that print to this window. If you want the code to be printed in the Session window then place an **echo** command before the code you want printed and a **noecho** command when you want to turn this off.

Macros can be nested; i.e. a macro may have in its body a statement of the form %`file` where `file` contains a macro.

## C.1.1 Control Statements

There are a number of statements that allow for control over the order of execution of Minitab commands in a macro.

### IF, ELSEIF, ELSE, ENDIF

The **if, elseif, clse, endif** command appears in the following structure

```
if expression1
block1
elseif expression2
block2
else
block3
endif
```

where *expression1* and *expression2* are logical expressions and *block1, block2* and *block3* are blocks of Minitab code. If *expression1* is true then *block1* is executed, if *expression1* is false and *expression2* is true then *block2* is executed and if both *expression1* and *expression2* are false then *block3* is executed. Note that if one of the expressions is a column of logical values then the expression evaluates as false if all entries are false and as true otherwise. There can be up to 50 **elseif** statements between **if** and **endif**. A logical expression is any expression involving comparison and logical operators that evaluates to true (1) or false (0). For example, the code

```
gmacro
uniform
random 100 c1;
uniform -1 1.
let k1=mean(c1)
if k1<-.5
let k2=0
elseif k1>-.5 and k1<0
```

```
let k2=1
elseif k1>0 and k1<.5
let k2=2
else
let k2=3
endif
print k2
endmacro
```

generates a sample of 100 from the uniform distribution on the interval $(-1, 1)$, computes the mean and outputs 0 if the mean is in $(-1, -.5)$, outputs 1 if the mean is in $(-.5, 0)$, outputs 2 if the mean is in $(0, .5)$ and outputs 3 if it is in $(.5, 1)$.

## DO, ENDDO

We saw an example of **do, enddo** in C.1. These statements appear in the following structure

```
do Ki = list
block
enddo
```

where *list* is a list of numbers, perhaps a patterned list such as $-8 : 8/2$, or stored constants. The Minitab code in *block* is executed for each value in the *list* with the constant Ki taking on that value. The numbers in the list can be in increasing or decreasing order.

## WHILE, ENDWHILE

The **while, endwhile** statements appear in the following structure

```
while expression
block
endwhile
```

where *expression* is a logical expression and the code in *block* is executed as long as *expression* is true. For example, the code

```
gmacro
stuff
random 2 c1;
```

```
    uniform 0 1.
    let k2=1
    let k1=c1(k2)
    while k1<.5
    let k2=k2+1
    if k1<2
    let k1=c1(k2)
    else
    break
    endif
    endwhile
    print k2
    endmacro
```

generates a sample of 2 from the $Uniform\,(0,1)$ and then finds the first value in the sample greater than .5 and prints its location in the sample and if no such value is found prints 3.

## NEXT

The **next** command can appear in a **do, enddo** or **while, endwhile** and passes control to the first statement after the **do** or **while**, whichever is relevant, and the loop variable is set to the next value in the list.

## BREAK

The **break** command can appear in a **do, enddo** or **while, endwhile** and passes control to the first statement after the **enddo** or **endwhile**, whichever is relevant.

## GOTO, MLABEL

The **goto** command allows the macro to skip over a number of statements in the file. This takes the following form

```
    goto V
      ⋮
    mlabel V
```

where the goto V statement passes control to the statement following mlabel V and V is a number.

## CALL, RETURN

Macros can be invoked from within macros by using statements of the form `%file`. This requires that the macros are in different files. In fact the macros can be in the same file, all having their own **gmacro** and **endmacro** statements and *templates*. When the file is invoked the first macro is processed. If the first macro needs to refer to the other macros in the file then this is done via the **call** and **return** commands. For example, suppose that a file contains two macros and the first macro needs to use the second one. Then this is implemented via the structure

    gmacro
    *template1*
    *body1*
    endmacro

    gmacro
    *template2*
    *body2*
    endmacro

where somewhere in *body1* there is the statement

    call *template2*

which transfers control from the first macro to the second macro and somewhere in *body2* there is the statement

    return

which returns control to the first macro.

## EXIT

The **exit** command stops the macro. A typical use would be as part of an **if, elseif, else, endif** where if a certain condition was satisfied no further statements in the macro are executed.

## PAUSE, RESUME

The **pause** command returns control to the Session window and session commands can then be invoked. Control is returned to the macro after a **resume** command is issued in the Session window.

## C.1.2  Startup Macro

You can place commands that you want to be executed every time you start or restart Minitab in a file called `startup.mac` in the default Minitab directory or the `macro` subdirectory of this directory. For example, you can use the **note** command in such a file to send yourself reminders or the **outfile** command if you always want to record your work in a particular file.

## C.1.3  Interactive Macros

A macro can write data to the Session window and accept input from the user. We have already discussed the **note** command which allows you to write comments to the Session window. The **write** command can be used to write the contents of columns and constants to the Session window. For example, the code

```
gmacro
stuff
random 100 c1;
uniform 0 1.
write "terminal" c1
endmacro
```

generates a sample of 100 from the Uniform(0,1) distribution into C1 and then writes this on the Session window. Of course we could also have accomplished this using the **print** command but recall that **write** allows for formatted output.

Input can be provided to a macro from the keyboard while the macro is running. This is carried out using the special file name `terminal` with the **read**, **set** or **insert** commands. For example, the code

```
note Read 10 observations into C1.
set c1;
file "terminal";
nobs=10.
```

prompts for 10 observations which are placed into C1. Note the use of the subcommands **file** and **nobs** to the **set** command. You can also use the **read**, **set** and **insert** commands in a macro with an **end** statement provided you place the data in the file as well. Also data can be read in from an external file but the name of the external file must be on the same line as the **file** subcommand and not on the same line as the command as in the session command and of course enclosed in single quotes.

The **yesno** command allows you to decide which commands you would like executed perhaps based on what the exec has already computed. For example, the code

note : Would you like to execute the macro random.txt?
yesno k1
if k1=1
%random.txt
endif

asks whether or not you wish to execute the macro in `random.txt`. If you answer **y** then K1 is given the value 1 and the macro `random.txt` is executed and if you answer **n** then K1 is given the value 0 and the macro `random.txt` is not executed.

## C.2   Local Macros

Local macros are more sophisticated than global macros. Basically all the features we have discussed for global macros can also be used in local macros. The major difference is that global macros operate only on the worksheet while local macros create temporary *local worksheets* which are used for computations without disturbing the *global worksheet*. The contents of local worksheets are not seen in the Session window. Also local macros can have arguments and subcommands. It is through arguments such as columns, constants, etc., which are passed to and passed out of the macro, that a local macro operates on the global worksheet. Subcommands to a local macro modify the behavior of the macro. Perhaps local macros are most useful when you want to create a truly new command in Minitab that behaves like the other commands we have been discussing throughout this manual. Because of their considerably more sophisticated nature we do not discuss local macros any further here and refer the reader to [2] in Appendix F.

## C.3   Execs

An exec is a file of Minitab commands that can be executed using the **execute** command. This is convenient if there is a set of commands that are repeatedly used as then we need only put them in a file and execute the file. Actually in Version 12 execs are still available but they have been superseded by global and local macros which have all the capabilities of execs and more. So if you are using Version 12 you can ignore this section and use macros exclusively.

## C.3.1   Creating and Using an Exec

You can use any editor to create a file of Minitab commands. Save the file as ASCII text. If you save the file with the file extension .mtb, then you can execute the file with the **execute** command by referring only to the file name while otherwise you must provide the file name and extension. If the file is not in the default Minitab directory, then you must provide the full pathname. Alternatively you can use the **journal** command, described in I.11.8, which places the commands you type in a file with the file extension .mtj. If you use this approach, then you must refer to the file name and extension.

For example, suppose the file random.txt in the default Minitab directory contains

```
echo
random 100 c1
let k1=mean(c1)
let k2=stdev(c1)
print k1 k2
histogram c1
```

then the command

```
MTB > execute 'random.txt' 3
Executing from file:  random.txt
```

causes the contents of random.txt to be executed three times. The **echo** command causes the commands in the file to be printed in the Session window each time they are executed and can be left out if this is not desirable. If you want only some of the commands to be printed then place the **echo** command before those you want printed and the **noecho** command before those you don't want printed. Any comments – lines preceded by # – will not be printed even after an **echo** command. If you want a comment printed then place the comment on a line with the **note** command.

The general syntax of the **execute** command is

**execute** 'file' V

where V indicates the number of times the commands in `file` are to be executed with V = 1 being the default. If V < 1, then the exec is not executed. This can be a useful feature when we want to conditionally execute the exec. For example, V could be the constant K1 which if it took the value 0 or less would cause the exec not to be executed.

It is also possible for one exec to call another. In fact there can be 5 levels of nested execs.

In the window environment the exec can be executed using F̲ile ▶ Other F̲iles ▶ R̲un an Exec.

## C.3.2   The CK Capability for Looping

A form of looping can be carried out using execs. We can denote a generic column by CKi and then use the constant Ki to iterate from column to column. For example, suppose that the worksheet has numeric data in columns C1-C100 and the file `stuff.txt` contains

let c101(k1)=mean(ck1)
add k1 1 k1

Then the commands

MTB > `let k1=1`
MTB > `execute 'stuff.txt' 100`

cause the mean of each of the columns C1-C100 to be computed and to be placed in the corresponding entry of C101. This kind of iteration can also be carried out for matrices (see Appendix D) using the MK capability.

It is possible to use the CK capability to deal with a variable number of columns. For example, suppose that you want to calculate the row means for the columns in a number of worksheets and the number of columns varies from worksheet to worksheet. Then the exec in the file `rowmeans.txt` with commands

let k2=k1+1
rmeans c1-ck1 ck2

can do this for each worksheet as in

```
MTB > let k1=30
MTB > execute 'rowmeans.txt'
Executing from file:  rowmeans.txt
```

where we specify that the worksheet has k1=30 columns and thus the row means are placed in C31.

## C.3.3   Interactive Execs

Input can be provided to an exec from the keyboard while the exec is running. This is carried out using the special file name **terminal** with the **read**, **set** or **insert** commands. For example, suppose the file input.txt contains

```
note Read 10 observations into C1.
set c1;
file "terminal";
nobs=10.
```

Then when it is executed we are prompted for 10 observations which are placed into C1.

The **yesno** command allows you to decide which commands you would like executed perhaps based on what the exec has already computed. For example, suppose that the file adjust.txt contains the commands

```
let k1=mean(c1)
print k1
note Would you like random.txt to be executed?
yesno k2
execute 'random.txt' k2
```

which calculate the mean of C1, print this out and then ask whether or not you wish to execute the exec in random.txt. It you answer y, then K2 is given the value 1 and random.txt is executed once. If you answer n, then K2 is given the value 0 and random.txt is not executed. So dependent upon what you see for the mean of C1 you can decide whether or not to execute the exec in random.txt.

## C.3.4   Startup Execs

You can place commands that you want to be executed every time you start or restart Minitab in a file called startup.mtb in the default Minitab directory.

For example, you can use the **note** command in such a file to send yourself reminders or the **outfile** command if you always want to record your work in a particular file.

# Appendix D

# Matrix Algebra in Minitab

Some versions of Minitab also have the facility for carrying out matrix algebra. This is useful sometimes as matrices can really simplify some complicated algebra and numerical work. In particular the computations associated with fitting the regression models discussed in II.10 and II.11 can be easily handled using matrix algebra. In this appendix we assume that you have been introduced to the basic operations and concepts of matrix algebra.

As an example consider fitting a quadratic polynomial $\beta_1 + \beta_2 x + \beta_3 x^2$ to $n$ data points $(x_1, y_1), \ldots, (x_n, y_n)$. To do this we must first create the matrices

$$X = \begin{pmatrix} 1 & x_1 & x_1^2 \\ \vdots & \vdots & \vdots \\ 1 & x_n & x_n^2 \end{pmatrix}$$

and

$$y = \begin{pmatrix} y_1 \\ \vdots \\ y_n \end{pmatrix}.$$

The matrix $X$ is called the *design matrix*. In a more advanced statistics course it is shown that best fitting quadratic (least-squares quadratic) is given by $b_1 + b_2 x + b_3 x^2$ where

$$b = \begin{pmatrix} b_1 \\ b_2 \\ b_3 \end{pmatrix} = \left(X^t X\right)^{-1} X^t y$$

the vector of predicted values is given by

$$\hat{y} = Xb$$

the residuals are given by

$$r = y - \hat{y}$$

and

$$s^2 = \frac{(y - \hat{y})^t (y - \hat{y})}{n - 3}$$

is the estimate of $\sigma^2$.

For the general linear model $E[y] = \beta_1 x_1 + \beta_2 x_2 + \cdots + \beta_k x_k$ where $x_1, \ldots, x_k$ are the explanatory variables and we observe the data $(y_i, x_{1i}, \ldots, x_{ki})$ for $i = 1, \ldots, n$. We present the data in matrix form as

$$X = \begin{pmatrix} x_{11} & \cdots & x_{1k} \\ \vdots & \vdots & \vdots \\ x_{n1} & \cdots & x_{nk} \end{pmatrix}$$

$$y = \begin{pmatrix} y_1 \\ \vdots \\ y_n \end{pmatrix}$$

Then the best fitting linear model is $b_1 x_1 + b_2 x_2 + \cdots + b_k x_k$ where

$$b = \left( X^t X \right)^{-1} X^t y$$

$$\hat{y} = Xb$$

$$r = y - \hat{y}$$

and

$$s^2 = \frac{(y - \hat{y})^t (y - \hat{y})}{n - k}$$

Notice that these formulas are the same no matter what linear model we are dealing with. There are many other useful quantities associated with the linear model that can be defined in terms of matrices.

# D.1  Creating Matrices

Matrices in Minitab are denoted by M1, M2, ...., M100. Note that there can be at most 100 matrices. The **name** command can be used to give alternative names to matrices. For example, the command

```
MTB > name m1 'design'
```

assigns the name **design** to the matrix M1 and it can be referred to as such afterwards with the name in single quotes.

If we are going to use matrices the first step is to create them. This can be done in a number of ways. For example, we can use the **read** command as in

```
MTB > read 5 3 m1
DATA> 1 1 1
DATA> 1 2 4
DATA> 1 3 9
DATA> 1 4 16
DATA> 1 5 25
  5 rows read.
MTB > print m1
  Matrix m1

  1 1  1
  1 2  4
  1 3  9
  1 4 16
  1 5 25
```

which creates a $5 \times 3$ matrix M1 for the fitting of a quadratic at the $x$ points 1, 2, 3, 4, 5. Note that the dimensions of the matrix accompany the **read** command with the number of rows followed by the number of columns and no **end** statement is required.

Sometimes you want a matrix with constant entries and the **define** command is available for this. The general syntax of this command is

**define** V $D_1$ $D_2$ $E_1$

which creates a matrix $E_1$ with $D_1$ rows, $D_2$ columns and every entry is the number V. Often you want to create a matrix with given entries along its

diagonal and 0's in all the off-diagonal elements. The **diagonal** command
with syntax

**diagonal** $E_1$ $E_2$

creates a square matrix $E_2$ with column $E_1$ in the diagonal and all other
entries 0. The matrix $E_2$ is square with dimension equal to the length of $E_1$.
If instead $E_1$ is a matrix and $E_2$ is a column then the diagonal of $E_1$ is placed
into the column $E_2$.

It is often convenient to copy the content of columns in a worksheet
directly into a matrix and conversely. For example, the commands

```
MTB > set c1
DATA> 5(1)
DATA> end
MTB > set c2
DATA> 1:5
DATA> end
MTB > let c3=c2*c2
MTB > copy c1 c2 c3 m1
```

create the matrix M1, printed above, fairly quickly using shortcut operations
associated with the **set** command and basic column operations. For large
patterned matrices this is probably the best way to create the matrix. Also if
the matrix is in an external file we can read the matrix into a set of columns
using the **read** command and then use the **copy** command to create the
matrix. If M1 is as above then the command

```
MTB > copy m1 c1-c3
```

copies the first column of M1 into C1, the second column of M1 into C2,
etc. Also we can create copies of matrices using the **copy** command. For
example,

```
MTB > copy m1 m2
```

creates a matrix M2 with the same entries as M1.

To delete matrices use the **erase** command. For example,

```
MTB > erase m1
```

deletes the matrix M1.

In the window environment general matrices can be created using Çalc ▶ Matrices ▶ Read, constant matrices using Çalc ▶ Matrices ▶ Define Constant, diagonal matrices using Çalc ▶ Matrices ▶ Diagonal, matrices can be copied from columns and vice versa using Çalc ▶ Matrices ▶ Copy and filling in the dialog boxes appropriately. In the window environment matrices can be directly read in from a file using Çalc ▶ Matrices ▶ Read and filling in the dialog box appropriately.

## D.2  Commands for Matrix Operations

In this section we give the syntax of the commands available in Minitab for matrix operations and provide an example.

**add** $E_1$ $E_2$ $E_3$ - puts $E_1 + E_2$ into $E_3$ where $E_1$, $E_2$, $E_3$ are matrices of the same dimension.

**eigen** $E_1$ $E_2$ $E_3$ - puts the eigenvalues of symmetric matrix $E_1$ into column $E_2$ and the eigenvectors into matrix $E_3$.

**invert** $E_1$ $E_2$ - puts $(E_1)^{-1}$ into $E_2$ for matrix $E_1$.

**multiply** $E_1$ $E_2$ $E_3$ - puts $E_1 E_2$ into $E_3$ where $E_1$ is a constant and $E_2$, $E_3$ are matrices of the same dimension or $E_1$ is a matrix with the same number of columns as the number of rows in matrix $E_2$.

**subtract** $E_1$ $E_2$ $E_3$ - puts $E_1 - E_2$ into $E_3$ where $E_1$, $E_2$, $E_3$ are matrices of the same dimension.

**transpose** $E_1$ $E_2$ - puts $(E_1)^t$ into $E_2$ for matrix $E_1$.

In the window environment these commands can be carried out using

Çalc ▶ Matrices ▶ Arithmetic
Çalc ▶ Matrices ▶ Eigen Analysis
Çalc ▶ Matrices ▶ Invert
Çalc ▶ Matrices ▶ Arithmetic
Çalc ▶ Matrices ▶ Arithmetic
Çalc ▶ Matrices ▶ Transpose

respectively and filling in the dialog boxes appropriately.

As an example suppose we consider fitting the least-squares quadratic when we have observed the data $(1, 7.2365)$, $(2, 17.2625)$, $(3, 33.6455)$, $(4, 55.4614)$ and $(5, 82.2756)$. We construct the $X$ matrix as M1 as indicated above and the $y$ values are placed in the matrix M2. Then the commands

```
MTB > transpose m1 m3
MTB > multiply m3 m1 m4
MTB > inverse m4 m4
MTB > multiply m4 m3 m5
MTB > multiply m5 m2 m6
MTB > print m6
 Matrix M6

 2.19780
 2.10946
 2.78638
```

compute the least-squares quadratic as $2.19780 + 2.10946x + 2.78638x^2$. The commands

```
MTB > multiply m1 m6 m7
MTB > subtract m2 m7 m8
MTB > transpose m8 m9
MTB > multiply m9 m8 m10
Answer = 0.1880
MTB > let k1=.1880/2
MTB > print m7 m8 k1
```

store the predicted values in M7, the residuals in M8 and $s^2$ in K1.

# Appendix E

# Advanced Statistical Methods in Minitab

While this manual has covered many of the features available in Minitab there are still a number of other useful techniques that, while not relevant to an elementary course, are extremely useful in a variety of contexts. The material in this manual is good preparation for using the more advanced techniques.

We list here the more advanced methods available in Version 12 and refer the reader to reference [3] in Appendix F for details.

- General linear model and general ANOVA

- Variance components

- MANOVA

- Principal components

- Factor analysis

- Discriminant analysis

- Cluster analysis

- Correspondence analysis

- Time series analysis – trend analysis, dDecomposition, etc.

- Exploratory data analysis

- Quality control tools – run chart, Pareto chart, etc.

- Reliability and survival analysis – accelerated life testing, probit analysis, etc.

- Design of experiments – factorial designs, response surface, mixture designs, etc.

# Appendix F

# References

[1 ] *Meet Minitab*, Release 12 (1997). Minitab Inc.

[2 ] *User's Guide 1: Data, Graphics, and Macros*, Release 12 (1997). Minitab Inc.

[3 ] *User's Guide 2: Data Analysis and Quality Tools* (1997). Minitab Inc.

# Index

212